Transport, Trade and Economic Growth – Coupled or Decoupled?

ifmo – Institut für Mobilitätsforschung (Ed.)

ifmo
Institut für Mobilitätsforschung (Ed.)
Eine Forschungseinrichtung der BMW Group

Transport, Trade and Economic Growth – Coupled or Decoupled?

An Inquiry into Relationships between Transport, Trade and Economic Growth and into User Preferences concerning Growth-oriented Transport Policy

With 6 Figures and 14 Tables

Springer

Editor
ifmo
Institut für Mobilitätsforschung
Eine Forschungseinrichtung der BMW Group
Kurfürstendamm 31
10719 Berlin
Germany
www.ifmo.de

ISBN 978-3-642-08791-2 e-ISBN 978-3-540-68299-8

Springer is a part of Springer Science+Business Media

springer.com

Cover: deblik, Berlin
Printed on acid-free paper 68/3180YL - 5 4 3 2 1 0

Foreword

Economics has traditionally been working with the abstraction of the point market in order to be able to focus on the processes of productivity growth and of general macroeconomic equilibrium. The insights of regional sciences on the one hand and of development economics on the other hand were generally ignored. Different methodical advances such as the treatment of economies of scale as internal to a model, respectively the question how a sustainable world economy could look like have lead economists to increasingly focus on questions of endogenous growth, of trade and therefore of transport. Transport quickly comes to the fore because the generalised costs of transport and travel still reduce the growth of the scale economies and because the external costs of transport significantly lower the welfare effects of that growth. The at the beginning of the 70ies outdated and inefficient regulation of the transport sector worldwide and the large differences in the success of regulatory regime change since then have further aroused interest. The visibly decreasing effects of further infrastructure expansion and a more than proportional growth in costs further increase the urgency of an efficient regulation. The fast growth of the internet as a means of trade and cooperation is also an expression of this predicament.

The present book, the revised dissertation thesis of the author, pertains to this discussion. After an in-depth account of the literature, the author shows what a significant role the quality of transport systems plays for the transport sector respectively for trade between two countries.

Despite all reductions in the generalised costs of transport since 1800, respectively 1950, accessibility still structures many exchange processes. The improved access to information about products, their prices and availability moves many markets, especially transport markets as air transportation shows, but distances retain their effects. In many respects we are still far away from ideal point markets! Further improvement of transportation systems, further reduction of the generalised costs of transport and its externalities, more efficient regulation of transport markets will therefore remain an important political and scientific task for a long time.

Zurich, February 2007

Kay Axhausen
Institute for Transport Planning and Transport Systems, ETH Zürich
Member of the Board of Trustees of the Institute for Mobility Research, Berlin

Dr. Walter Hell
Director of the Institute for Mobility Research, Berlin

Preface

The division of labour is the strongest promoter of the rising productivity of an economy, probably even stronger than technological progress. This holds true for functional as well as for regional division of labour. It impressively becomes manifest in the fact that wherever trade barriers are eliminated an intensive exchange of goods and recently also services sets in. There is a reason to why it holds true almost with the stability of a law of nature that transport volume between two developed economies is rising at a rate 1.5 fold of national income growth. This process has definitely not come to an end and is permanently given new impetus through the expansion of the EU, the liberalisation of the world trade regime and globalisation in general. The efficient exchange of goods and the therefore essential transport capacities are thus a strategic factor for the conservation of welfare in industrialised countries.

In the past trade flows were restricted by protectionist measures of single nations or by the lack of financial resources for the expansion of the transport network, but today, at least in Europe, worries concerning the sustainable supply with adequate energy sources and concerns over the external costs of transport play an important role. The present study takes on this challenge and puts the quantitative transport necessities together with qualitative needs of a modern economy into a framework that opens new perspectives for an efficient transport policy.

The starting point for these considerations is the fact that the capacity of important terrestrial transport systems cannot be expanded endlessly due to political, social and economic reasons. Thus, transport capacity is scarce. This basic assumption is accompanied by the following central facts:

- The international specialization is evolving dynamically and this development does not seem to end any soon. This induces overproportional transport flows with regard to economic growth.
- If this transport demand meets with constraints in the transport network, on the one hand transport costs indirectly increase (longer transporting times) while on the other hand the quality of transport services (reliability, calculability) decreases.
- This indirect increase in costs and the loss of quality lower the attractiveness of transport and therefore limit international specialisation. Especially goods requiring high quality transport services are affected.

It suggests itself that the crowding out of high quality transport does not make much sense economically due to the fact that it is especially this kind of transport that has the potential to stimulate productivity. This is why the control mechanism described above should urgently be replaced by a different one. By the way, sectoral bans on freighters that at first glance seem to eliminate only "inferior" transports are for sure inefficient: erroneously it is assumed that expensive goods should be transported while cheap goods should not be transported. But in fact goods should be transported if the trading partner's comparative advantage compared to his customers is especially large. This may well lead to transporting bananas being more economically efficient than the transport of basic industrial products. The only possibility to force people to reveal their preferences is by means of price regulation. The fact that transport volume reacts very sensitively to price changes – with an elasticity of – 2 % – is also nicely shown in the present study. With a product being offered on a competitive market one can easily lean back and leave everything up to the market. But transport network services are not traded on competitive markets, they are at best natural monopolies with only imperfectly endogenised external effects and therefore virtually predestined for "political" price setting.

The present book shows a way out exactly from this unsatisfying dilemma between a dysfunctional market and a political price mandate little substantiated by basing the central question – What is an efficient price for transport net services? – on a new notion of efficiency. Assuming a fixed transport net the price is optimal if the net welfare loss (thus, only the dead weight loss of the price increase) of lower transport volumes can be compensated by the welfare gains due to a better transport quality. This leads us again to the marginal principle well known in economics. The precondition for this approach is the evaluation of transport quality in monetary terms. The conjoint-analysis that is applied offers a methodically attractive as well as an approved approach.

The applied methods of course can and should be improved and developed further but with the present study a prototype has been produced that opens up doors for new research approaches. Results though especially nurture hopes that on this imperfect market of transport net services quasi-market efficiency can be reached without having to renounce to the necessary political framing.

Prof. Dr. Gottfried Tappeiner
Department of Economics
Leopold-Franzens-Universität Innsbruck

Contents

List of Figures

List of Tables

Abbreviations

ACA	Adaptive Conjoint Analysis
CA	Conjoint Analysis
CIF	Cost, Insurance, Freight
CLS	Cluster
DESTATIS	German National Statistics Institute
FOB	Free on Board
GDP	Gross Domestic Product
ISTAT	Italian National Statistics Institute
NCT	Neoclassical Theory
NTT	New Trade Theory
OECD	Organization for Economic Cooperation and Development
OLS	Ordinary Least Squares
PDF	Production Function
SITC	Standard International Trade Classification
TC	Transport Costs
tkm	Tonnes-Kilometers
TV	Trade Volume

Management Summary (German)

Güterverkehr, Handel und Wirtschaftswachstum – Gekoppelt oder Entkoppelt?

Eine Analyse der Beziehungen zwischen Güterverkehr, Handel und Wirtschaftswachstum sowie der Nutzerpräferenzen von Unternehmen hinsichtlich einer wachstumsorientierten Verkehrspolitik

Ist zusätzliches Wirtschaftswachstum untrennbar mit zusätzlichem Güterverkehrswachstum verbunden?

Politische Entscheidungsträger und Unternehmer sind mit einer ähnlichen Herausforderung konfrontiert: Alle Möglichkeiten zur Förderung des Wirtschaftswachstums und zur Erschließung neuer Märkte sollten in Zeiten des internationalen Standortwettbewerbs genutzt werden. Ein wesentlicher Bestandteil einer wachstumsorientierten Strategie liegt in der Bereitstellung sowie effizienten Nutzung einer entsprechenden Verkehrsinfrastruktur. Andererseits jedoch ist eine zunehmende Skepsis gegenüber zusätzlichen Verkehrsinfrastrukturinvestitionen in der öffentlichen Diskussion festzustellen. Vor allem hinsichtlich des Ausbaus der Straßenverkehrsinfrastruktur und dem zunehmenden Güterverkehrswachstum bestehen Vorbehalte. So gerät auch die Transportindustrie verstärkt unter Druck aufgrund ökologischer Bedenken. In Anbetracht dieses Dilemmas stellt sich die Frage, ob zusätzliches Wirtschaftswachstum untrennbar mit zusätzlichem Güterverkehrswachstum verbunden ist, wie dies bisher angenommen wird.

Zur Beantwortung dieser Frage erfolgt im ersten Teil des Buches eine Analyse der Beziehung zwischen Wirtschaftsleistung und Güterverkehr. Bisherige Untersuchungen konzentrieren sich dabei hauptsächlich auf den Einfluss von zusätzlichen Infrastrukturinvestitionen auf das Wirtschaftswachstum.[1] Diese empirischen Arbeiten weisen mittels Produktions- oder Kostenfunktionen größtenteils einen positiven und signifikanten Einfluss von zusätzlichen Infrastruktur-

1 Aschauer, D. A.; (1989a; b; 1991); Bougheas, S.; et al.; (1997)

investitionen auf das Bruttoinlandsprodukt nach. Dabei werden allerdings die zugrunde liegenden Mechanismen, wie Wirtschaftsleistung und Güterverkehr zusammenhängen, nicht untersucht. So kann davon ausgegangen werden, dass Transport eine effizientere Organisation der Produktion ermöglicht, was wiederum zu einer erhöhten Wirtschaftsleistung führt. Durch den Transport von Ressourcen, Halb- und Fertigprodukten kann dort produziert werden, wo dies aus ökonomischen Gründen sinnvoll ist. Skaleneffekte als auch Faktorpreisunterschiede können ausgenutzt werden. Wird tatsächlich mehr transportiert, um Skaleneffekte und Faktorpreisunterschiede auszunutzen, spiegeln Handelsvolumina dies wider. Betrachtet man die Auswirkungen von Veränderungen bilateraler Handelsvolumina auf die Wirtschaftsleistung kann so der positive Beitrag des Transports abgeschätzt werden. Neben der Betrachtung der Wirkungen von Handel auf die Wirtschaftsleistung, muss jedoch auch die Beziehung zwischen Handel und den Transportbedingungen – vor allem der Transportkosten – betrachtet werden, um den Zusammenhang zwischen Transport, Handel und Wirtschaftswachstum umfassend zu untersuchen.

Analyse der Auswirkungen von Veränderung bilateraler Handelsvolumina auf die Wirtschaftsleistung

Transportelastizität bilateraler Handelsströme entspricht -2

In der vorliegenden Arbeit werden sowohl der Zusammenhang zwischen Transportkosten und bilateralen Handelsvolumina als auch die Wirkungen von Veränderungen der Handelsvolumina auf die Wirtschaftsleistung untersucht. Dabei ergab eine empirische Analyse bilateraler Handelsströme auf regionaler Ebene zwischen Deutschland und Italien eine Transportkostenelastizität von –2 für den Zeitraum von 1993 bis 2003. So lässt sich für den Zusammenhang zwischen Transportkosten und Handelsvolumen feststellen: Wenn die Transportkosten um 1 % steigen, sinkt in Folge das bilaterale Handelsvolumen um 2 %. Handelsströme reagieren also äußerst sensitiv auf Veränderungen der Transportkosten.

Um festzustellen, welche Auswirkungen sich durch diese Veränderungen auf die Wirtschaftsleistung ergeben, erfolgt im nächsten Schritt eine Analyse der Beziehung zwischen den exportierten Gütermengen und der regionalen Wirtschaftsleistung. Dabei deuten die Ergebnisse auf einen signifikant positiven Zusammenhang zwischen den beiden Größen hin: Der Handelsanteil ist eine wichtige Determinante der Wirtschaftsleistung.

Mit dieser Vorgehensweise – der Orientierung auf das Handelsvolumen als Proxy für die tatsächlich stattfindende Güterverkehrsleistung – wird die bisherige Form der Analyse wesentlich erweitert. Dadurch wird es möglich, die Ergebnisse der vorangegangenen empirischen Arbeiten in einen größeren Zusammenhang zu stellen und zu interpretieren. Man kann den bisher implizit angenommenen positiven Zusammenhang zwischen tatsächlich stattfindendem Güterverkehr und der Wirtschaftsleistung detaillierter erfassen.

Neben der makroökonomischen Betrachtung erfolgt eine Analyse der Präferenzen der Wirtschaftssubjekte

Um neben gesamtwirtschaftlich interessanten Ergebnissen einen Beitrag zu verkehrspolitischen Überlegungen zu liefern, scheint es neben der makroökonomischen Betrachtungsweise auch wichtig, näher zu analysieren, welche Präferenzen die Wirtschaftssubjekte haben. Da über die Präferenzen einzelner Unternehmen aus der verarbeitenden Industrie sowie Handelsunternehmen bisher wenig bekannt ist, wird im zweiten Teil des Buches der Frage nachgegangen, welche Attribute für Logistiker den tatsächlichen Nutzen der Transportleistung bestimmen. Dazu wird eine Methode aus dem Bereich der Konsumforschung angewandt. Mit Hilfe einer adaptiven Conjoint-Analyse werden die Einschätzungen von Logistikern erhoben und ausgewertet, um so auf Optimierungspotentiale schließen zu können.

Die Qualität der Güterverkehrsleistung spielt eine ähnlich bedeutende Rolle wie die Transportkosten

Diese Analyse zeigt, dass für Unternehmen neben den überaus bedeutenden Transportkosten, gerade die Qualität des Güterverkehrs eine große Rolle spielt. Der rechtzeitige und zuverlässige Transport, also die Qualität der Transportleistung, ist wichtig. So wichtig, dass Qualitätskomponenten unter Umständen sogar Kostenvorteile kompensieren können. Diese Feststellung ist vor allem im Hinblick auf politische Entscheidungen über Verkehrsinfrastrukturen und deren Nutzung interessant. Verkehrspolitische Überlegungen sollten sich nicht nur auf die Höhe der Investitionen in Verkehrsinfrastrukturen beschränken, sondern auch die Qualität und die für die Nutzer entstehenden Vorteile der verfügbaren Verkehrsinfrastrukturen einbeziehen. Genauso sollten auch Regulierungen des Zugangs zu Verkehrsinfrastruktur stets aus Nutzerpräferenzen abgeleitet werden.

Die Ergebnisse der vorliegenden Analyse, in der unterschiedliche Methoden kombiniert werden, um eine andere Perspektive auf die bestehende Kopplung zwischen Güterverkehrsleistung und Wirtschaftsleistung zu erreichen, legen nahe, dass:

1. das Transportvolumen sehr sensitiv auf Veränderungen der Transportkosten reagiert;
2. die Verfügbarkeit von Transportvolumina einen deutlich positiven Einfluss auf das regionale Wirtschaftswachstum zu haben scheint;

3. aus der Sicht der Logistiker die Transportleistung als Kombination von Qualität und Kosten zu beurteilen ist;
4. die Qualität der Transportleistung in vielen Aspekten, bei gegebenem Infrastrukturvolumen, eng mit dem Verkehrsaufkommen verknüpft ist;
5. Qualitätsverbesserungen unter bestimmten Bedingungen sogar monetäre Kostenvorteile kompensieren können.

Mit dieser Analyse werden erste konkrete Hinweise auf das Verhältnis von Güterverkehr, dem Handelsvolumen und dem Wirtschaftswachstum, sowie den Präferenzen von Unternehmen geliefert und damit Fingerzeige für die Ausrichtung einer wachstumsorientierten Verkehrspolitik gegeben. Aufbauend auf den Ergebnissen der vorliegenden Analyse können sich weitere Forschungsarbeiten mit der Klärung von Optimierungspotentialen von Verkehrsinfrastrukturen auseinandersetzen, um Imperative für eine wachstumsorientierte Verkehrspolitik aufzuzeigen.

1 Freight Transport and Economic Growth – Two Inseparable Twins?

The way economic interactions take place is changing rapidly. Decreasing trade barriers and increasing ambitions of managers have led to a new perception of economic and geographic space, resulting in a massive increase in world trade over the past decades. GATT/WTO rounds have caused a substantial reduction of tariff and quota regulations while physical trade barriers have been steadily decreasing due to declining transport costs fostered by technological improvements in transport and logistics. Looking at the numbers, from 1948 to 1998 world merchandize trade grew at 6 % annually, and has therefore increased three times faster than GDP growing at an average annual rate of 1,9 % over the same period of time.[2]

Massive increases of world merchandize trade

Specialization and concentration in production, growing separation of stages in value-creation as well as the increasing internationalization of economic activities of companies lead to growing transport volumes as well as to increases in average transportation distances. Historically, economic growth has gone hand in hand with increases in freight transport. Over the past decades the value-added process has become even more transport intensive. Between 1985 and 1998 freight transport performance measured in tkm increased by 54 % in the EU, while GDP grew by 35 %.[3] From 1970 to 2000 road freight intensity, measured in tonnes-kilometers/Euro of GDP, in the EU 15 countries has increased by 1 % p.a.[4]

Increases in freight transport intensity of GDP

Though the numbers seem to be telling straightforward a certain story, in the political debate other imperatives are set by multiple stakeholders. Due to rising public concerns over emission and noise pollution from freight transport ad-

2 WTO, (2007)
3 Gilbert, R.; Nadeau, K.; (2001)
4 European Commission, Directorate General, Joint Research Centre; (2003b)

Decoupling is a key issue in the EU White Paper on Transport Policy

ditional increases of freight transport are getting more and more contentious. For example, the EU White Paper on Transport Policy from 2001 identified the decoupling of economic growth from freight transport growth as one of the key issues to be tackled on the way towards more sustainable transport.[5] According to this proposition, freight transport growth is to be reduced without limiting economic growth. In order to achieve this, the relations between these two factors need to be scrutinized thoroughly.

The present study aims at analyzing the structural relationship between freight transport and economic development in detail in order to provide deeper insights into the coupling between these two factors. In this context two main aspects are considered:

It is important to scrutinize the link between GDP and freight transport

1. Since one is not able to deduce a form of causality from the empirical correlation between economic growth and freight transport it is important to further explore the underlying mechanisms linking these two factors. Simply because GDP and freight transport show similar patterns of growth, this does not necessarily imply that there is an underlying causal relationship. Thus, in order to be able to provide further information on the link between these two factors it seems to be essential to examine their relationship in theoretical and empirical models.
2. Due to the fact that large-scale expansion of existing transport infrastructure networks in the European Union are regarded as rather contentious and given the indication that there exists a close link between transport, trade and economic growth, it is of special importance that infrastructure capacity is directed towards its most efficient use. Therefore, it is necessary to gain insight into user preferences concerning transport services.

Gaining insight into user preferences

Firstly, the analysis concentrates on the nature of the relationship between economic development and freight transport growth and thus on the decoupling potential. Secondly, it aims at providing insight into the effects of changes in transport conditions on the economy in order to be able to estimate the impacts of transport policy measures influencing transport conditions.

Previous studies on the relationship between transport and economic growth were mainly concerned with the impact of

5 European Commission; (2001)

(further) infrastructure investment on output (growth). In doing so, they focused on the "input" variable investment instead of looking at the transportation actually happening on roads, rail, by sea and in the air. Thus, the direct link between an input variable and economic growth has been scrutinized. Opposed to this, the present study aspires to analyze the underlying mechanisms behind the relationship between transport and economic growth more thoroughly. In this analysis the connection between transport and output is not a direct one but rather intertwined with its influence on production structures and processes, location and plant size decisions, distribution structures and processes and other characteristics of industrial organization. Trade determines the degree of specialization, of economies of scale, of exploitation of comparative advantage, of competition and of generation and diffusion of technological progress. With that form of scrutiny of the relationship between transport costs and trade and subsequently between trade and output, the present study examines the impact of changes in transport conditions on output.

Analyzing the underlying mechanisms behind the relationships between freight transport and economic growth more thoroughly

The results of this study widely confirm theoretical propositions and empirical evidence in the literature. Therefore, it is essential to enhance the efficiency of the transport sector also by providing an adequate transport infrastructure in order to foster trade and economic growth. However, it is not appropriate to take into account only transport costs, as this is often done. Rather, it is essential to consider qualitative aspects explicitly, especially as they tend to gain in importance due to production processes becoming more complex and production techniques like just-in-time or just-in-sequence being adopted on a larger scale.

Lower trade barriers and better transport conditions do foster economic growth

Since there seems to be a substantial lack of in-depth studies of the demand side[6] and since political decision-makers and business people have to use the existing transportation network in an efficient way given the public concerns over further infrastructure investment and given that transport still seems to be very important for promoting economic well-being in developed economies, it seems to be worthwhile to come up with a combination of two methods. This

6 Holtz-Eakin (1994) also underlines the need for microeconomic research due to the lacks of macroeconomic models: "Because there likely are narrow circumstances in which the productivity effects are positive, future research in this area should be devoted to making more precise the microeconomic linkage between the provision of infrastructure and the nature of the production process."

The combination of two analytical approaches

combination of two analytical approaches is important for two reasons: First, the macroeconomic perspective provides us with further insight into the relationships at the aggregate level while second, the microeconomic approach allows us to find out more about the desiderata of the users of the existing transport infrastructure, or more specifically, examines the actual needs and wants of commercial users with respect to the transport system.

Following that reasoning, the major aim for the second part of the present study is to quantify the relative importance commercial users of the transport system attach on qualitative aspects of transport services, such as speed or reliability of punctual arrival versus monetary costs. This was done by surveying logistic departments of manufacturing and distribution companies with an adaptive conjoint analysis to get deeper insights into their preferences concerning different aspects of transport quality. This allows determining which characteristics of transport services matter most for commercial users, and therefore will facilitate to design transport policy measures that help to make the best possible use of the given infrastructure.

The present study is organized as follows: chapter 2.1 summarizes the existing literature focusing on the impact of infrastructure investment on output (growth) as well as on studies on the relationship between transport costs and trade, between trade and economic growth. Additionally, studies on the location decision, production structure and economic growth are examined.

An econometric analysis between transport costs and trade volume and between trade volume and GDP at the regional level

The empirical part starts in chapter 2.2 with an econometric analysis of the relation between changes in transport costs and trade volume, which subsequently affects economic growth. Chapter 2.2.1 applies gravity equations to analyze the effect of transport cost on trade volume on a regional level. The relation between trade volume and GDP growth is studied empirically in chapter 2.2.2.

2 Coupled or Decoupled? An Inquiry into the Relationships between Transport, Trade and Economic Growth

2.1 Literature Review

The literature review is subdivided into two strands of literature. One, which is directly focusing on the impact of changes in infrastructure investment on output growth. In the other strand of literature scholars have been analyzing the processes that are caused by the possibility to transport goods and subsequently lead to welfare gains. Chapter 2.1.1 consists of studies modeling production and cost functions, whereas the independent variable is aggregate public infrastructure investment and the dependent variable is represented by aggregated output growth.

The second part of the literature review then analyzes the processes induced by the possibility to transport goods. It focuses on the impact of transport on trade and subsequently on the impact of trade on economic growth. A separate part is dedicated to transport and location decisions. Chapter 2.1.2 discusses the relation between transport costs and trade. In chapter 2.1.3 an examination of the gains from trade links trade with economic growth. In chapter 2.1.4 literature on the linkage between transportation related aspects and location decisions is revisited.

2.1.1 Transport Infrastructure Investment and Economic Growth

Historically, the link between economic development and transport infrastructure investment is widely recognized. At the beginning of the industrialization process, the possibility to transport goods was a necessary precondition for specialization and the rise of the manufacturing industry. As long as investment is devoted to establish basic links in order to be able to transport at all, the necessity of infrastructure investment is not questioned. So in developing countries even today infrastructure investment mainly serves to build a basic network of links as infrastructure supply is still at a very low level.

The impact of further transport infrastructure investment is controversial

But in developed countries which are in most cases endowed with a well-connected transport infrastructure, the economic impact of further infrastructure investment is controversial.

The law of decreasing marginal returns

According to the law of decreasing marginal returns to investment, additional infrastructure investment in economies already well-endowed with transport infrastructure, is supposed to lead to lower rates of return. However, special characteristics of transport infrastructure may render the applicability of this theory to transport infrastructure investment difficult.

Network structure of transport infrastructure

Due to the network structure of transport infrastructures, the establishment of a new link may well lead to substantial returns as the impact of one particular link may be felt throughout the whole network. This holds especially true for bottlenecks, where the speed of flow may be drastically reduced due to a sudden capacity reduction, while the rest of the network's capacity may still be underutilized.

"One of the lessons that can be learned from production function modeling is that ongoing improvements of transport infrastructure will result in a lower growth of the regional value added. However, if the other production factors – capital and labor – show a growing tendency there might occurr a bottleneck if the level of transport infrastructure remains constant. This bottleneck may have a negative impact on the productivity of labor and capital." [7]

Congestion

This leads to a second feature of transport infrastructures which makes the concept of decreasing returns to investment disputable, namely the problem of congestion. Increasing transport demand renders networks with unchanged or with only insufficiently adjusted capacity congested. Thus, an increase in demand leads to a situation comparable to the status before the investment. This is due to the fact, that an investment that is used for capacity expansion induces an increase in capacity but this will be used up as soon as demand increases at least by the same amount as the capacity is expanded. For developed economies with high transport demand this means that, due to congestion, the law of decreasing marginal returns can not be applied to transport

7 Bruinsma, F.; (1995), p. 10

infrastructure as availability of service does not increase although total capital stock increases.[8]

Economic and demographic characteristics

Thirdly, the degree to which infrastructure investment impacts on economic growth strongly depends on the economic and demographic characteristics of the region where the investment takes place. E. g. different economic actors will react differently to changes in accessibility, different housing situations (e.g. single versus other households) will lead to differences in impacts of infrastructure investment. Thus, the nature of the local economy and the different actors that make decisions influence the magnitude of the impact of infrastructure investment on economic growth.[9]

Summing up, the necessity of further infrastructure investment in developed countries which are endowed with a rather good transport infrastructure supply is controversial i. a. due to the law of decreasing returns to investment. However, several characteristics of transport infrastructure and of its usage imply that it may very well be conducive to economic growth to invest in further transport infrastructure in developed countries.

In order to be able **to measure the impact of transport infrastructure investment on economic growth**, a theoretical framework has to be established which specifies the links between these two factors.

Firstly, **transport infrastructure investments lead to changes in the relative prices of accessibility of various locations**. These price changes necessarily alter the relative advantage of spatially located activities and thus the economic opportunities. This is due to the fact that costs of inputs and the prices of outputs in different locations vary with costs of accessibility in these locations.

Secondly, changes in infrastructure supply lead to changes in transport economic behavior. In the short to medium term, trip generation rates, travel volumes or choice of routes are altered, while long term changes result in changes in location decisions of households and firms and in changes in land and property prices.

Thirdly, these changes are subsequently translated into measurable economic benefits such as improvements in factor productivity, larger output, increased demand for inputs, increased property values and greater demand for consumer goods.[10]

8 Bougheas et al.; (1997)
9 Banister, D.; Berechman, Y.; (2000)
10 Banister, D.; Berechman, Y.; (2000)

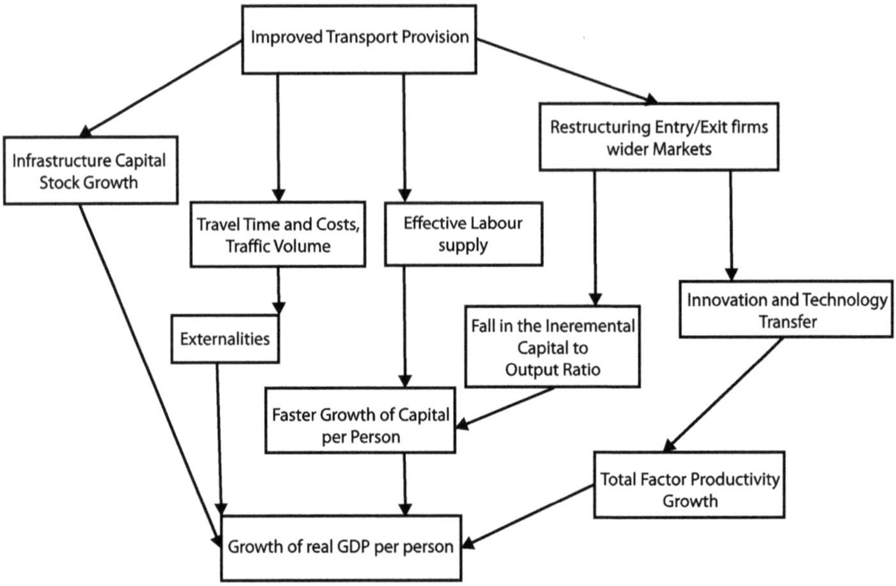

Fig. 2.1 Improvement of Transport Provision and Economic Growth
Source: OECD Papers 2003b

Distance is a constraint to economic activity

Distance can be regarded as a constraint to economic activity. Improvements in the transport system lessen this constraint,

"thereby enabling the attainment of higher economic development. ... If transport can be regarded as a constraint on the attainment of economic opportunities in an area, then it can be regarded as a necessary condition." [11]

Figure 2.1 schematically shows some of the above depicted effects.

The controversial issue of the measurement of the economic impact of transport infrastructure investment has been subject of various economic research studies. Basically, in economic research two different modeling approaches are used to analyze the economic impact of transport infrastructure investment: The production function approach and the cost function approach.

11 Banister, D.; Berechman, Y.; (2000), p. 37

2.1.1.1 Production Function Modeling

The neoclassical growth model

In the traditional neoclassical growth model (Solow, 1956), growth is explained in terms of the availability and the use of productive factor inputs. Concerning infrastructure, this means that its availability increases the productivity of physical and human capital, which leads to lower production costs. Furthermore, infrastructure serves as a direct factor input and thus, assuming positive marginal returns to investment, leads to an increase in output.[12]

The positive effect of public capital investment on private sector output

The public capital hypothesis, which was developed in the framework of the neoclassical growth model, states that there is a positive effect of public capital investment on private sector output.[13]

"Public infrastructure investment is regarded as the trigger mechanism to the increasing of private capital rates of return through increases in private capital stock and labor productivity, which in turn result in higher total output and economic growth".[14]

There are two ways of how public infrastructure investment can be incorporated in the production function:

1. Public infrastructure investment is introduced in the PDF as a private factor input, which, under the assumption of positive returns to investment, leads to an increase in output. As these factor inputs are assumed to be provided costless, private sector productivity increases.[15]

   $$Y = f(C, L) = C^{\alpha} L^{\beta} \qquad 2.1$$

   C denotes capital inputs, L labor inputs, Y is output. As C increases due to increases in the public capital stock, private sector output increases as well, without occurring additional cost on the private sector, thus productivity increases.

12 Guild, R. L.; (1998)
13 Arrow, K. J.; Kurz, M.; (1970)
14 Banister, D.; Berechman, J.; (2000)
15 This poses a major restriction of these models, as public infrastructure investment is financed by taxes and thus increases the tax rate, which not only represents the actual costs of the investment but which leads to market distortions and thus to welfare losses.

2. Another strand of literature models transport infrastructure investment not as an additional factor input but similar to technical progress, thus raising the overall level of the PDF and consequently the level of output.

$$Y = f(C, AL) = AC^{\alpha} L^{\beta} \qquad 2.2$$

A direct and an indirect effect of aggregate public capital on output growth

A denotes technological progress and raises the overall level of the PDF.

Aschauer (1989b), who examined the relationship between aggregate public capital and output growth, discovered a direct as well as an indirect effect. There is a direct effect because public capital acts as an intermediary. It is introduced as a direct input in private production and thus raises private sector output, as it is assumed that the marginal productivity of public capital is positive. Concerning the indirect effect, infrastructure does not serve as a direct input factor but is supposed to raise the marginal productivity of private factor inputs which is due to the assumption of complementarity between public and private capital investment. This means that an increase in the public capital stock raises the marginal productivity of private capital inputs. Due to the higher marginal productivity of private inputs, the demand for private inputs rises. So the higher marginal product of private capital raises private capital formation and thus leads to a rise in private sector output. [16]

Empirical Studies Using Production Functions

The decline in public infrastructure investment was a causal factor of the US productivity slowdown in the 70ies and 80ies

Aschauer (1989a) found so strong an economic impact of infrastructure investment that the productivity slowdown in the US in the 70ies and 80ies may be attributed to a decrease in public infrastructure investment due to the above stated positive effect of public capital on private output. By way of production function modeling, his empirical investigation identified the decline in public infrastructure investment to be a causal factor of the US productivity slowdown. Munnell (1990a) confirms this finding and states that

"the drop in labor productivity has not been due to some mystical concept of multifactor productivity or technical progress.

16 Aschauer, D. A.; (1989b)

Rather, it has been due to a decline in the growth of public infrastructure".

Aschauer's (1989a) findings are significant enough to state that the marginal productivity of public investment is higher than from private capital. Output elasticities of public capital are between 0.38 – 0.56. The specification used is a Cobb-Douglas production function. The data used are national US aggregate output data for the time between 1949 and 1985. Non-military public capital is employed to model public capital stock, whereas he adds that core infrastructure such as highways, airports and sewers is more productive than any other type of infrastructure. McGuire (1992) and Deno (1988) confirm the finding that highway capital is more productive than other types of infrastructure capital, whereas Morrison and Schwartz (1996) only include highways, water and sewers because they state that the estimated impact of public capital on costs of production in the manufacturing sector "is somewhat smaller if we include other public capital". Capacity utilization and time dummies are used to control for the influence of business cycles and time effects.[17]

The marginal productivity of public investment is higher than from private capital

Highway capital is more productive than other types of infrastructure capital

Aschauer (1989b) reached similar conclusions using panel data for the Group Seven countries. His study produces elasticities between 0.34 – 0.73. Munnell (1990b) produced similar results for the time between 1948 and 1987, using a Cobb-Douglas specification, national US aggregate output data and aggregate public capital investment. Output elasticities of public capital are positive and significant between 0.34 – 0.37. In a subsequent study Munnell (1990a) uses a translog Cobb-Douglas specification for state-level data. Elasticities are still positive and significant at 0.16. The use of panel data controls for invariant state- and time-specific effects. Garcia-Milà and McGuire (1992) and Eberts (1986) also used panel data for the US and found positive and significant, if smaller, output elasticities for public capital.

Spatial spill-overs between states and regions are the reason for the difference in results depending on the level of aggregation of data.[18] Concerning the aggregation level of data the use of state-level data may lead to an underestimation of the productivity of public capital because it cannot capture the aggregate effects of public capital in a system.

The use of state-level data may underestimate the productivity of public capital

17 Kim, J. H.; (2003)

18 Cohen, J. P.; Morrison, C. J.; (2001)

Boarnet (1996) distinguishes a direct and an indirect effect. The direct effect denotes the economic impact within the same jurisdiction that contains the infrastructure investment. The indirect effect arises outside of the particular jurisdiction where the public infrastructure capital is invested. Indirect effects can be positive as well as negative. If infrastructure investment results in relocation decisions positive effects in some regions are accompanied by decreasing economic activity in other regions. If the former are not offset by the later, public capital creates positive cross-state spillovers. Thus, overall indirect effects from public capital are positive which means that an infrastructure investment in one state leads to benefits in other states.[19] The overall effect of infrastructure investment is positive if the indirect effect is either positive or negative but smaller than the direct effect. Boarnet (1996) argues that if infrastructure is productive at all, the overall effect is likely to be positive.

"If transportation is productive for counties (as the positive coefficient on own county street and highway capital suggests), then theoretically the direct and indirect effects ought not perfectly cancel. Even if all increased output in one region is due to the production of firms which moved into the county, the direct effect of increased output in the region must be larger than the indirect effect of decreased output in the other counties where the migrant firms were previously located."

This is due to the fact that firms which relocated are producing more productively than previously because the possibility of higher productivity was the reason for firms to relocate in the first place.

If it is assumed that the overall effect is positive, the relatively high elasticities between public infrastructure and aggregate private sector output can partly be explained by the high level of aggregation of data used in the empirical estimation. Aschauer (1989a,b) uses aggregate national time-series data, while other studies work with regional or local data (Garcia-Milà and McGuire, 1992; Hulten and Schwab 1991). Garcia-Milà and McGuire (1992) still found positive links between infrastructure investment and output. Hulten and Schwab (1984), Eberts (1986), Eberts and Fogarty (1987) and Duffy-Deno and Eberts (1991) examined the link between metropolitan infrastructure investment, output, per-

19 Munnell, A. H.; (1992)

sonal income and private investment and found a positive relationship. They tested for the direction of causation and found that causality runs mostly from infrastructure capital to output growth.

Causality runs mostly from infrastructure capital to output (growth)

Bougheas et al. (1997) do not treat infrastructure as an input in the production of final goods but rather model infrastructure "as a technology which reduces costs in the production of intermediate inputs and, therefore, fosters specialization" and competition, which leads to higher output. This is due to the fact that both competition and specialization are impeded by fixed costs. The empirical results show that the effects of infrastructure on output and economic growth are, with few exceptions, positive.

Stephan (2001) focuses on the regional level of the "Bundesländer" and treats infrastructure as an unpaid factor of private production, hence it enters the production function of private firms. He finds a significant positive effect of public capital on private manufacturing for all tested econometric models and specifications. Whereas differences in public capital intensity can not explain differences in observed levels of output, differences in changes of public capital can explain differences in changes of output. This correlation holds in the short- as well as in the long-run. But diagnostic tests reveal that the specification issues such as serial correlation, groupwise heteroscedasticity and cross-sectional correlation are present.

Summing up, the **majority of the studies produces significant and positive elasticities** between **public infrastructure capital and output**, whereas numbers are smaller from studies using disaggregate data than from those using aggregate data. So it can be concluded, "that there are considerable productivity advantages associated with public infrastructures".[20] Conrad and Seitz (1997) also emphasize the positive effect of public infrastructure capital on private capital formation.

There are considerable productivity advantages associated with public infrastructures

However, concerning the type of data with respect to the size of the area under analysis, international comparison of results can be misleading. For example, a US state can often be considered the equivalent of a European country, and a German state can be similar in scale to the whole of Greece.[21]

International comparison of results can be misleading

20 Conrad, K.; Seitz, H., (1997), p. 716

21 Rivolis, A; Spence, N., (2002)

Specification problems

Apart from this, various specification problems threaten the plausibility of the positive results. Some are specifically due to the use of the production function approach and can thus be eliminated by the use of cost functions whereas others also arise when cost functions are applied. Problems immanent to the production function approach are examined in the next section, the cost function approach and its problems are explained in more detail in chapter 2.1.1.2.

Methodological Problems of the Production Function Approach

Consideration of allocative inefficiency

Under the traditional growth accounting approach employing production function modeling it is assumed that factors are paid their marginal products. If this does not hold true firms may be producing inefficiently and thus below the frontier of production technology. In this case, total factor productivity may rise due to improved efficiency and not only due to technological progress as is assumed by traditional growth theory. Applying a dual framework using short-run variable cost functions allows for allocative inefficiency.[22]

The effect of factor input prices on factor utilization

When production functions are estimated, firms' optimization decisions with respect to how much output to produce and what mix of inputs to use in the production process is not considered. So in order to be able to consider firms' optimization, the marginal productivity conditions for the inputs should be estimated jointly with the production function. The production function approach omits the effect of factor input prices on factor utilization. So production function models do not consider the interactions between infrastructure capital and internal and external economies of scale which lead to cost reductions in private production and thus affect the firms' demand for factor inputs. Most studies using production function models usually disregard the potentially large impact of public infrastructure on firms' costs and thus on economic performance, even when fixity effects are incorporated. **Unless the firms' optimization conditions are included in the estimation, results are likely to be seriously mismeasured due to the omission of the impact of public infrastructure on firms' costs.** Opposed to this, the cost function approach provides cost elasticities of output as

22 Boisso, D.; et al.; (2000)

well as specific effects of infrastructure capital on demand for private sector input.[23]

A particular shortcoming of the Cobb-Douglas specification is the restriction it imposes on the elasticity of substitution of the input factors. Thus, it specifies the degree of economies of scale, which means that it is determined a-priori whether it is a function with constant, increasing or decreasing economies of scale.

Restriction of the Cobb-Douglas specification

2.1.1.2 Cost Function Modeling

Using the cost function approach, input quantities and production costs are assumed to be endogenous, the level of output and input prices is supposed to be exogenous.

The studies using cost functions widely differ from each other which is due to the use of diverse data sets, assumptions about the optimization behavior of the firms and the specification of the cost function. Most studies use either translog or generalized Leontief functional forms, which means that the proportion of the factor inputs used varies with output.

Apart from this, different notions of public capital are used, just like in the studies using the production function approach. Some focus on core infrastructure such as highways, bridges or sewers, while others use the total stock of public capital.

Despite these differences, most studies using the cost function approach reach the conclusion that public capital contributes positively to productivity by generating cost savings.

Public capital contributes positively to productivity by generating cost savings

Empirical Studies Using Cost Functions

Lynde and Richmond (1992) employ aggregate US nonfinancial corporate business sector data for the period from 1958 to 1989 employing a translog cost function. They impose the restriction of constant returns to scale on all inputs and assume perfect competition.

Nadiri and Mamuneas (1994) also use a translog cost function to estimate private production cost reductions from public capital for 12 manufacturing industries in the U.S. from 1955 to 1986. Both studies find that public infrastruc-

23 Morrison, C. J., Schwartz, A. E.; (1992)

ture capital substantially reduces private production costs.[24] Nadiri and Mamuneas get a **net return of 35 %** for Highway Capital for the 1950s and 1960s which is attributed to the fact that the Interstate was being built. The net effects were cumulative.[25]

In another study Lynde and Richmond (1993) use a translog cost function to estimate output elasticity of public capital for U.K. manufacturing. They incorporate quarterly data for the period 1966 – 1990 in the estimation and come to the conclusion that 40 percent of the U.K.' s observed productivity slowdown can be attributed to the decline in the public capital to manufacturing labor ratio.[26]

Morrison and Schwartz (1992) employ state-level data for the manufacturing sectors in the U.S. for the years from 1970 to 1987 in a variable cost function. They use a Leontief specification, which means that the proportion of the factor inputs used varies with output. Public and private capital variables are treated as exogenous, whereas public capital is a factor explaining observed scale effects. The cost effects of infrastructure, fixity and internal economies of scale can be reflected by the cost elasticity with respect to output derived from the variable cost function. External scale economies stemming from outside forces of public capital will cause output and total cost changes to be non-proportional. **Summing up, scale effects may arise from short run fixities of inputs, scale economies imbedded in the technology, or external impacts from public infrastructure.** Investment in highways, water facilities and sewers was used to represent public infrastructure capital.[27]

Berndt and Hanson (1992) incorporate aggregate data from the Swedish private sector into a short-run variable cost function under the assumption that private and public capital are fixed in the short run. They too reach the conclusion that public capital reduces private costs.

Conrad and Seitz (1994) estimating a translog cost function for the manufacturing, construction and trade, and transport sectors of West Germany for the period 1960 – 1988 also find substantial cost reductions in these sectors due to infrastructure investment.[28]

24 FHWA, (1998)
25 Lakshmanan, T. R.; Anderson, W. P.; (2002)
26 FHWA, (1998)
27 Morrison, C. J.; Schwartz, A. E.; (1992)
28 FHWA, (1998)

Haughwout (2002), using a spatial equilibrium model of an economy with non-traded, localized public goods like infrastructure, includes household preferences in his analysis.

"The key insight is that infrastructure's effect on firms and households is mediated through its effect on local equilibrium prices. While the productivity of public capital is certainly one component of its aggregate social value, many of the most prominent authors in the ongoing debate over infrastructure productivity have recognized that direct, non-pecuniary, household benefits are a second avenue by which infrastructure may affect welfare."

In his analysis he comes to the conclusion that exogenously given infrastructure is envisioned as a contributor to local property values, consequentially raising local income.

Summing up, using a cost function approach allows consideration of the impact of factor prices on factor utilization and therefore on private sector input. Opposed to this, in production function analysis production inputs are assumed to be exogenous.

Studies using cost functions usually produce **a positive contribution of infrastructure to output growth** although its **magnitude** seems to be **lower than when production functions are used**. Concerning the type of infrastructure it should be noted that some types, e.g. highway capital, are more productive than others.

2.1.1.3 (Specification) Problems Concerning Both Approaches

Spurious Correlation Due to Non-stationarity of the Data

Common trends in data

The existence of a strong correlation between public infrastructure investment and productivity growth does not necessarily reflect a causal relationship. It may as well result from a spurious correlation between production and public capital because both labor productivity and public infrastructure spending have declined over the same period of time. But this decline in both factors can also be attributed to other forces being causal. Common trends in data may lead to this kind of spurious correlation.

"For example, if two wholly unrelated measures have similar time trends, they can exhibit an apparent, statistically significant relationship between themselves when, in fact, no economic relationship exists."[29]

Thus, non-stationarity in aggregate time series data may be responsible for spurious correlation between infrastructure capital and output growth.[30]

"In time series analysis, estimating a relationship between non-stationary variables that are not co-integrated gives rise to the problem of spurious regression; the error term in the regression is non stationary, producing a high degree of 'noise' in the relationship, and inconsistent parameter estimates."[31]

If non-stationarity is ignored, ordinary least squares in levels may produce statistically significant coefficient estimates when in fact there is no relationship. Usually the problem of non-stationarity is eliminated by transforming the data to first differences. But by taking first differences of the variables, short-run capital accumulation to short-run changes in output is estimated whereas "long-term relationships in the data are destroyed and therefore it does not make economic sense to use equations in this form".[32] This argument renders Aschauer's (1989a,b) and Munnell's results more plausible as they do not take first differences of the variables.

Using panel data to cope with spurious correlation

Another way of coping with spurious correlation is by using panel data. The use of panel data over several states lowers the possibility of spurious correlation as national trends vary over states and therefore common trends in the data are less likely to occur. Most studies using state-level panel data still get positive and significant elasticities of public capital on private sector output although most results are substantially lower than for national aggregates.[33]

Hulten and Schwab (1984) test for spurious correlation and come to the conclusion that the differences in output growth are not due to differences in the growth of public infrastructure but rather to variation in the rates of growth of capital and labor.

29 Stephan, A.; (1997)
30 Duggal, V. G. et al.; (1999)
31 Canning, D.; (1989)
32 Duggal, V. G. et al.; (1999)
33 FHWA; (1998)

Reverse Causation

In case of reverse causation the positive coefficient for public capital could well be the effect of productivity growth on infrastructure capital rather than the reverse. Due to reverse causation, the error terms and the capital stocks may be correlated, which makes it unclear whether the contribution of capital to output or the effect of output on capital accumulation is estimated.

Microeconomic Considerations

The approaches depicted above linking transport infrastructure investment and economic growth use aggregated macroeconomic data. They are supposed to capture the benefits that actually occur at the micro level in an aggregated setting, but the mechanisms operating at the micro level that lead to these welfare gains are not explored in detail.

Cost Benefit Analysis to assess microeconomic benefits

A conventional tool to assess microeconomic benefits is the Cost Benefit Analysis. It captures the increase in consumer surplus due to a downward shift of the supply function. Due to the fact that the CBA is a static analysis, dynamic gains cannot be captured, for example a shift in the demand curve. Other factors not included are represented by location effects, which include agglomeration economies such as urbanization and localization economies.[34]

Apart from this, changes in the supply chain are not included as well:

"The overall changes throughout the supply chain may even require going beyond the usual partial equilibrium view of cost-benefit analysis which is based on the presumption that the relative prices of goods remain largely unchanged. Far-reaching transport policy measures or large transport infrastructure investment projects are likely to be associated with changes in relative prices which can induce changes in the geographical pattern of trade relations and a relocation of production activities. The exclusion of such "secondary effects" *from the evaluation of transport policy projects implies a tendency to underestimate the net benefits of transport policy".*[35]

34 Gains due to optimization of location decision are explored in more detail in part 4 of chapter 1.

35 ECMT, (2003), p. 3

Probably the most severe lack of microeconomic analysis is the fact that the degree of market extension is exogenous and thus independent from transport infrastructure. So welfare gains due to the improvement of transport infrastructure which enables firms to reach broader markets are not included in CBA.[36]

2.1.1.4 Conclusions

Previous research is based on the link between transport infrastructure investment and output (growth)

Previous research on the relationship between transport and economic growth is usually based on the relationship between transport infrastructure investment and output (growth). This link is examined by production or cost function modeling using aggregated data on transport infrastructure investment and aggregated output data. The following three equations show how public infrastructure investment can impact output, the first two use PDFs, the third is a cost function.

Using production functions, the impact of infrastructure investment on output is usually modeled in either of the following two ways:

1. Public infrastructure investment is introduced in the PDF as if it was a private factor input and, under the assumption of positive returns to investment it raises output.

$$Y = f(C, L) = C^{\alpha} L^{\beta} \qquad 2.3$$

Thus, by increasing C (capital), Y increases as well. It is assumed that public infrastructure investment is provided costless, thus private industry productivity necessarily increases.

2. Public infrastructure investment can be modeled similar to technological progress, thus raising the whole PDF to a higher level.

$$Y = f(C, AL) = AC^{\alpha} L^{\beta} \qquad 2.4$$

3. Using cost functions, output and input prices are exogenously determined while input quantities and production costs should be minimized. In these studies, the effect of

36 Lakshmanan, T. R.; Anderson, W. P.; (2002)

public infrastructure investment is reflected in a change in production costs and not in output.

$$D = f(w, Y) \qquad 2.5$$

L... denotes labor
C... denotes capital
Y... denotes output
w... denotes input prices
D... denotes costs (D for dollars)

Both approaches, the production as well as the cost function approach, produce results showing productivity gains induced by infrastructure investment. Due to differences in the modeling procedures the magnitude of productivity gains varies. Although more recent research produces smaller elasticities of public capital on private sector output than Auschauer (1989), results are still positive and significant.

Positive impact of public capital on private sector output

The general **conclusion** that can be drawn from previous research on the link between public infrastructure investment and private sector output is that there is **a positive impact of public capital on private sector output** and **thus on economic growth** though the **magnitude** of this impact is **not clear**.

It is also shown that **public infrastructure investment leads to higher private capital formation.**

Specification problems cast doubt on the results

Specification problems cast doubt on the plausibility of the results. Especially **spurious correlation** and **reverse causation** may raise doubts concerning the obtained results.

The fact that studies produce positive and significant results does not necessarily lead to the conclusion that further infrastructure investment is efficient. Under the assumption that production factors are not working to full capacity, investment ALWAYS leads to output growth, no matter whether people are employed to dig wholes into the ground or to build roads. As soon as wages are paid, regardless of the nature of the work that is done, GDP increases. This holds true even if the utility for other branches of the economy or for individuals is questionable. Only if infrastructure investment increases the productivity of private factor inputs can benefits in other branches be reaped.

Tautological relationship between investment and output growth

Apart from this, the fact that an increase in investment leads to an increase in output is a relationship that may be called **tautological**. The derivation of positive and significant results does not allow us to automatically draw the conclusion that further infrastructure investment is economically efficient, which depends on the magnitude of the effect of infrastructure investment on aggregate output growth. Aschauer (1997a) for example shows that

"there is a non-linear relationship between public capital and economic growth such that permanent increases in the public capital ratio bring forth permanent increases in growth – but only if the marginal product of public capital exceeds the after-tax marginal product of private capital." [37]

There is not sufficient evidence that further infrastructure investment is economically efficient

Thus, significant and positive results for the impact of public infrastructure investment on output without comparing them to the marginal product of private capital does not provide sufficient evidence for drawing the conclusion that further infrastructure investment is economically efficient.

Although some authors question the high estimates of Aschauer's findings, they do not regard this as evidence "to support the argument that infrastructure is unproductive. Rather, they suggest that the inability of the literature to identify the benefits of infrastructure stems from the use of aggregate data".[38] Holtz-Eakin (1994) emphasizes the need for an approach digging deeper:

"... future research in this area should be devoted to making more precise the microeconomic linkage between the provision of infrastructure and the nature of the production process." [39]

The impact on output growth itself can be regarded as the outcome of various processes such as internal economies of scale, external economies of agglomeration and urbanization induced by infrastructure investments. These underlying causes that lead to the examined gains in output growth do not receive much attention in the previous literature studying the economic impact of transport.

Summing up, **the microeconomic approach** performed at the firm level allowing a more detailed analysis **is inclined to underestimate the true welfare gains** due to its limited

37 Aschauer (1997a) shows that for most of the US during the 70s and 80s the actual level of public capital were below the level which would have maximized the rate of productivity growth.

38 Bougheas, S. et al.; (1999), p. 169

39 Holtz-Eakin, D.; (1994), p. 20

scope, while **the macroeconomic approach** by means of production or cost function modeling **does not examine the underlying mechanisms leading to welfare gains**. So the obtained results, which vary in magnitude, cannot be underpinned by evidence of the processes leading to these results. Thus, the criticism of the existing macroeconomic research on the relationship between transport and economic growth might also be due to the fact that it is performed by measuring outcomes instead of analyzing the underlying causes.

Outcomes are studied instead of analyzing the underlying mechanisms

Now, I could endeavor on producing another macroeconomic study directly analyzing the impact of transport infrastructure investment on output trying to get rid off the specification problems but, especially in the case where results are quite controversial, it seems to make more sense to **examine the underlying causes** in order to be able to better estimate impacts and directions of these impacts. Although the analysis is still of the macroeconomic kind it is performed on a lower and more detailed level in the sense that not only input – output linkages are examined but also the underlying mechanisms as for example trade flows which lead to higher output via the possibility to transport goods in a more efficient way.

There is a vast amount of literature analyzing the processes that are caused by the possibility to transport goods, and their economic impacts. Due to the fact that international trade enabled by the possibility to transport goods leads to significant welfare gains, trade theory represents an important source of literature examining the processes caused by transportation and subsequently leading to welfare gains.[40]

Trade theory examines the processes leading to welfare gains

The subsequent part of the literature review examines economic processes that are induced by changes in transport infrastructure provision and that consequently lead to changes in output. In chapter 2.1.2 the relationship between transport and trade, while in chapter 2.1.3 the link between trade and output growth is examined. Still, some aspects of productivity gains due to improved transport infrastructure can not or at least not entirely be captured within the scope of an analysis of trade flows. Part of these effects are explored in chapter 2.1.4 on the relationship between transport, location decision and production structure.

40 At this point I would like to mention that trade on a national basis leads to substantial welfare gains as well. But due to measurement difficulties this mechanism can not taken into account.

2.1.2 Transport Costs and Trade

Distance acts as a barrier to trade

An important mechanism linking transport to productivity gains is trade.[41] Distance between trading partners represents a natural barrier to trade. Ceteris paribus, the higher the distance between trading partners, trade theory as well as empirical data show that the lower will be the resulting trade volumes between these partners. By enabling to overcome distance, transport makes trade possible in the first place. Still, by transporting goods from one place to the other, resources, in the form of transport costs, are consumed that could otherwise be diverted to other uses.

Transport costs are part of overall trade costs also comprising tariffs, administration costs etc. Hence, their impact on trade can be compared to other impediments to trade and influences trade volume in a similar way as other trade costs. Of course, ceteris paribus, **the lower the trade costs**, the lower the barriers to trade and **the higher the resulting trade volumes**.

"A fall in trade costs has two effects on the volume of trade. First, for given relative factor endowments, it raises the degree of specialization of countries, leading to a larger volume of trade in the short run. Second, it raises (lowers) the factor price of each country's abundant (scarce) production factor, leading to diverging paths of relative factor endowments across countries and a rising degree of specialization over time. This creates an additional effect on the future volume of trade that adds to the static and dynamic effects of future reductions in trade costs."[42]

Transport infrastructure investment is assumed to cause decreases in transport costs which again lead to higher trade volumes. These are supposed to provoke substantial welfare gains.

Differences in transport costs account for differences in international competitiveness

"Variations in transport costs across countries may be able to account for differences in their ability to compete in international markets. Thus, differences in the volume and quality of infrastructure across countries may be responsible for differ-

41 Due to measurement difficulties in this study only international trade will be taken into account.

42 Cunat, A.; Maffezzoli, M.; (2003), p. 2

ences in transport costs which in turn, may be able to account for differences in competitiveness."[43]

Low transport costs – a necessary but not a sufficient condition

Still, a fall in trade costs alone is not a sufficient condition for the development of a higher degree of specialization. By lowering the barriers to trade specialization is fostered but not made possible in the first place, hence the fall in trade costs is a necessary but of course not a sufficient condition for the promotion of specialization. So in order for trade to occur in the first place, trade costs need to be below a certain threshold. **The efficiency benefits of trade must supersede the cost of interregional shipment and other trade costs.**[44]

Trade induced welfare impacts

Concerning trade-induced welfare impacts, it can be distinguished between the terms-of-trade effect and the resource cost effect. The terms-of-trade effect refers to the increase in traded goods prices due to transport costs, which means that the slope of the supply curve for traded goods increases. The resource cost effect denotes impacts caused by the consumption of resources in the transport sector which could otherwise be used for the production of traded goods, hence decreasing production of traded goods. If transport costs are reduced both effects lead to positive impacts on welfare. Whether both effects can be taken into account depends on how transport costs are measured.

2.1.2.1 The Measurement of Transport Costs

Four different ways of measuring transport costs can be distinguished: Uniform ad-valorem iceberg costs, distance and geography related proxies, transport costs based on CIF/FOB value of imports and exports, and real freight expenditures.[45]

Samuelson (1952) developed the **iceberg cost approach**, whereas it is assumed that **transport** is **not a separate sector** but an "implicit consumption of the commodity being transported"[46], which means that transport costs are conceived as an impediment consuming resources differentially between products. Thus, the rate of consumption depends on distance as well as on the product transported. Interactions with other

43 Bougheas, S. et al. ; (1997), p. 170
44 Lakshmanan, T. R. ; Anderson, W. P. ; (2002)
45 Combes, P.-P.; Lafourcade, M.; (2003)
46 Vickerman, R.; (1995), p. 229

factors of the production process are also taken into account as e.g. the degree of scale economies impacts on the importance of transport costs.

Not considering transport as a separate sector leads to the fact that resources used for transport do not decrease the available resources for production. So, costs for transport can only be modeled as the price wedge between the domestic and the foreign good. Other **impacts** are **not taken into account** which **may lead to** their **underestimation.**[47]

Uniform ad-valorem iceberg approach: transport costs are assumed to be linear

Under the **uniform ad-valorem iceberg approach**, transport costs are assumed to be proportional to commodity prices. This leads to transport costs being linear. Both assumptions are unrealistic, Hummels (2002) finds that freight rates are not linear in the value of the goods transported.

Although empirical evidence has shown that **distance** is a good **proxy to measure transport costs**, several factors influencing transport costs are **not taken into account** such as **infrastructure quality** or **real geography**. This approach could be substantially improved by including real geography and infrastructure variables. Apart from this, studies using distances often apply the "great circle" formula, which means that instead of using existing routes to measure distances, longitude and latitude of the capital or "economic centre" are used. Hence, measures underestimate real distances goods need to travel.[48]

Transport costs measured as distance between CIF and FOB prices

As a substantial amount of the literature examining the economic impacts of transport costs concentrates on their impacts on trade, usually the **difference between CIF and FOB** prices are used **to measure transport costs** (Baier and Bergstrand, 2001; Limao and Venables, 2001; Radelet and Sachs, 1998). This means that the costs of transporting the goods from the production site to the port of exit and from the port of entry to the point of final distribution are not included despite land transportation often accounts for a substantial amount of transport costs as per km costs are on average 35 respectively 6 times higher for road respectively for rail than for long-distance sea transport.[49] Another fact that may lead to the **CIF/FOB measure underestimating actual transport costs** is selection bias as "high transport cost routes may systematically involve goods among which

47 Steininger, K. W.; (2001)
48 Head, K.; (2000)
49 Crook, G.; (2002)

transport costs are the lowest",[50] which lowers the CIF/FOB measure which represents an average over all commodities.

The real value of transport costs is often underestimated

Summing up, as the difference between CIF and FOB prices is frequently used to measure transport costs,[51] their real value is often underestimated. This leads to their impact on trade being underestimated too, but still they "prove to be one of the best measures available" [52], as they are much more exact than proxies for distance or iceberg transport costs.

2.1.2.2 *Transport Costs as Restrictions to Trade*

The impact of transport costs on trade is comparable to tariffs

Studies examining the impact of transport costs and tariffs on trade come to the conclusion that their magnitudes are at least comparable.

Waters (1970) and Finger and Yeats (1976) study US import data from the mid 1960s. Sampson and Yeats (1977) and Conlon (1982) examine Australian import and export data from the early 1970s. **All authors agree** on the fact that **transport costs pose a barrier similar in magnitude, or larger, than tariffs**. The studies of Yi (2003), finding US tariffs in 1994 to be 4.5 per cent of goods value, and Hummel (1999), finding US import weighted transportation costs to be 3.9 per cent in the same year, confirm this result. In addition Hummel finds these costs to be an important determinant of bilateral variation of trade.

Baier and Bergstrand (2001) found differing results by disentangling the relative impacts of income growth, income convergence, tariff-reductions and transport cost declines on the growth of international trade for a selection of 16 OECD countries. Their results show that GDP growth represents the most significant cause for trade growth with around 70 % , tariff-reductions being the second largest impact with a little more than 20 % and only around 8 % of trade growth can be explained by transport cost reductions, while GDP convergence has virtually no impact. From this, the conclusion might be drawn that transport cost reductions can be regarded as a minor influencing factor on the development of international trade. The study presents empirical evidence of an analysis of 16 OECD countries between the late 50s

50 Combes, P.-P.; Lafourcade, M.; (2003), p. 8

51 e. g. in Harrigan (1993), Gallup et al. (1999), Baier and Bergstrand (2001)

52 Combes, P.-P.; Lafourcade, M.; (2003), p. 8

and the late 80s, which means that changes in transport costs are regarded relative to the existing level of transport costs. Consequently, the conclusion that for these particular levels of transport costs in the analyzed countries within this time frame transport cost reductions have not been a major incentive for the growth of international trade could be drawn. But this does not allow the general conclusion of transport costs not being significantly influential on the development of trade as the existing level of transport costs mainly impacts on the influence of alterations in transport costs. It can be assumed that at an already low level of transport costs further reductions will not have significant impacts on the development of trade. Thus, what is important is the existing level of transport costs.

Apart from the above depicted difficulties concerning the measurement of transport costs, there is a **substantial difference between the weighted costs of transportation and the mean value of transport costs**. Mazzenga and Ravn, for example, (2002) find the mean of the costs across goods at the 4 digit level to be 10.1 per cent for 1994, Hummels (1998) finds similar number in 2 digit SITC data for the US and the IMF typically uses an estimate of 11 per cent. According to Mazzenga and Ravn (2002) the mean of the 4 digit costs of transportation falls from 14.4 percent in 1974 to 10.1 in 1994 whereas the weighted costs of transportation decreases from 6.31 percent to 3.49 percent. They attribute the relatively higher decrease in weighted transport costs to a possible substitution towards imports of goods with lower costs.[53] In addition, they add that the mean of the costs of transportation across individual goods types being much higher than the modal value of theses costs might be

"evidence that the true costs of transportation may be much higher than the measured costs of transportation in the sense that there is little or no trade in goods with high costs of transportation".[54]

There is a substantial substitution away from goods with high transport costs

Hummels (1998) who examined this in more detail came to a similar conclusion, hence, that there is substantial substitution away from goods with high costs of transportation.

A major part of the studies examining the causes of the rapid growth of world trade shows that the increase in the

53 Transport costs are measured as the difference between CIF and FOB import prices.

54 Mazzenga, E.; Ravn, M.; (2002), p. 5

export share of output is to be attributed to tariff barrier reductions. But in order to be able to draw this conclusion unrealistically high price-import elasticities need to be assumed. However, if transportation costs are included when studying trade expansion due to falling trade costs, trade expands more for a given elasticity. Bridgman (2003) finds that

"trade expansion due to falling trade costs is 43.4 percent higher when falling transportation costs are included."

Bridgman shows that the impact of changes in transport costs on trade patterns, in particular on the average length of haul, are different depending on the weight/value ratio of the shipped goods. The length of haul of low value goods is much more sensitive to transport price changes than of high value goods.

Concerning the welfare impacts of a reduction in the costs of transportation Mazzenga and Ravn (2002) find that a

"drop in the costs of transportation from 20 to 15 percent is equivalent to a permanent increase in consumption of just above 1.5 percent".[55]

They attribute the large welfare effects of transport cost reductions to the assumption of countries being specialized in production.[56] It is emphasized that the magnitude of the welfare effects is relatively independent of the value of the elasticity of substitution between the domestic and the foreign good. This finding is attributed to the fact that a low elasticity of substitution makes foreign goods less dispensable and, hence,

"although the effects of transaction costs on trade shares are lower when domestic and foreign goods are less substitutable, the effect of changes in trade shares on welfare are more dramatic when this elasticity is low".[57]

Apart from this, they allowed for home-bias, otherwise "the welfare effects would have been even larger", but they did not

55 Mazzenga, E.; Ravn, M.; (2002), p. 12

56 The assumption of transport costs being 20 percent might be unrealistically high and therefore the impacts of the reduction may be overestimated since it can be assumed that welfare impacts of reductions are higher the higher the initial level of transport costs is.

57 Mazzenga, E.; Ravn, M.; (2002), p. 12

allow for congestion effects in international trade, which, if included, would have led to smaller welfare effects.

Bond (1997) examines the welfare effects of transport cost reductions in a simple partial equilibrium model in which there are two traded goods. He states two favorable effects of transport cost reductions on welfare:

"*1. terms of trade for the small importing country are improved by reducing the costs of imports and raising the return per unit of exports (at given prices in the rest of the world,*
2. it has a favorable effect on the volume of imports. The favorable terms of trade effect is proportional to the volume of trade, indicating that reductions in transport costs are more beneficial when the volume of trade with the rest of the world is large." [58]

2.1.2.3 Conclusions

Transport costs act similarly as trade barriers

Studies examining the surge of world trade often attribute it to tariff barrier reductions. But their findings only hold true if unrealistically high elasticities of substitution between home and foreign good are assumed. If transport cost reductions are included in the analysis, much lower elasticities of substitution can be adopted. This has two effects: on the one hand, the results gain plausibility, and on the other hand, welfare impacts of trade tend to be higher when elasticities of substitution are low.

The importance of transport costs is frequently underestimated

From the literature **it can be concluded** that concerning the impact of transport costs on trade, **transport costs pose a similar restriction to trade as tariffs**. Furthermore, as trade barriers are continuously lowered, **the importance of transport costs as a determinant of the volume of trade flows is steadily increasing.**[59] Apart from this, transport costs are frequently underestimated which leads to an underestimation of their importance as well. Despite the fact that the importance of transport costs is estimated to be about equal to the importance of tariff barriers, the repercussions of changes in transport costs on trade have been given much less attention in research than tariff barrier reductions.[60]

Although transport costs impact trade similarly as tariffs, they have to be modeled differently as they usually are not

58 Bond, E. W.; (1997), p. 6
59 Micco, A.; Pérez, N.; (2001)
60 Combes, P.-P.; Lafourcade, M.; (2003)

proportional to output value but to distance (up to a certain amount), which vice versa usually is not proportional to the amount of border crossings.

2.1.3 Trade and Economic Growth

Linking transport costs to economic growth through trade

After the examination of the relationship between transport costs and trade, I proceed with the analysis of the interlinkage between trade and economic growth in order to be able to combine these two direct connections into one indirect relationship between transport costs and economic growth via trade volume.

Before starting with the examination of the relationship between trade and economic growth I would like to give a brief overview of the theoretical concepts of economic growth and how trade fits in.

2.1.3.1 Economic Growth

The neoclassical model of economic growth

The central assumption of the neoclassical model of economic growth, which is widely referred to under the name of its orginators, the **Solow-Swan model**, is the falling productivity of capital which in the end leads to zero per capita growth. In the Solow-Swan model the only variable influencing long-term economic growth is **technological progress** which is assumed to be **exogenous to the model**. This means that the **steady-state growth rate is fully determined by population growth and exogenous technological progress.** Concerning the rate of public investment and its consequences for economic growth, this means that

"... an increase in spending on productive capital will induce a period of temporarily high investment, but the pace of capital accumulation, and of economic growth will slow over time as the accumulation of capital diminishes the return to capital and the incentive for further investment. In the long run, the level of output will be higher but the growth rate of output will return to the same level as before the public spending."[61]

Thus, as it lacks an endogenous component explaining long-term economic growth, the **Solow-model of economic**

61 Aschauer, D. A.; (1997a), p. 3

growth only predicts short-term growth which represents a major restriction of this theory.[62]

Including technological progress as an endogenous variable in a neoclassical model poses substantial difficulties as the assumption of technical progress leads to increasing returns to scale. This is due to the fact that technological progress is partly a non-rivalry good which tends to result in increasing returns to scale. The later are inconsistent with the central assumption of perfect competition in the neoclassical model. In order to **effectively predict long-term per capita economic growth** the restrictive assumption that it is determined by the exogenous variable of technological progress has to be substituted by **models using endogenous variables as determinants of long-term economic growth**. Endogenous growth theory

"provides the tools to handle endogenous technological change and innovation within a dynamic equilibrium framework setting. This allows us to develop tractable and flexible models that embody the vision of economic life as an endless succession of innovation and change wrought by competition. With these tools we can bring to bear all ... about incentives, organizations and institutions, not only on the problem of economic growth per se but also on the many other economic phenomena that interact with growth."[63]

Technological progress needs to be included when modelling trade in a long-term growth model

As the sole determinant of long-run economic growth in the neo-classical model is represented by exogenous technological progress, interactions with other countries have no effect on an economy's long-term growth rate. In order to be able **to model trade in a long-term growth model, technological progress has to be included** in the model. Both **Romer** and **Schumpeter** emphasize the **beneficial role of trade for growth as it increases market size**, so the possibility for successful innovators to appropriate **monopoly rents increases**.

The **Frankel-Romer model** of endogenous growth

"is characterized by a scale effect, whereby the larger the number of firms in the economy, the larger the amount of knowledge externalities and therefore the higher the growth rate. Insofar as international trade increases market size and raises the equilibrium number of firms, it will be responsible for a rise in the economy's rate of growth. ... Second, it enlarges the

62 Barro, R. J.; Sala-i-Martin, X.; (1995)

63 Markusen, J. R.; (1995), p. 3

scope for knowledge spillovers, both of which are conducive to faster technological change."[64]

Basic differences between exogenous and endogenous growth

The basic difference between models of exogenous and endogenous growth is the existence respectively the absence of decreasing capital productivities. The absence of decreasing capital productivities basically explains the endogeneity of growth.[65] The endogenous growth approach provides "a fertile analytical ground for building models where international trade might affect the long run growth rate of the economy"[66].

Assuming **small open economies**, the possibility to **evade diminishing returns to capital** is made possible **by factor price equalization**. According to Markusen et al. (1995) the factor-price-equalization theorem states that

"Under identical constant-returns-to-scale production technologies, free trade in commodities will equalize relative factor prices through the equalization of relative commodity prices, so long as both countries produce both goods."[67]

Diminishing returns to capital can be avoided by allowing for free trade

Free trade allows a country that accumulates capital to avoid diminishing returns to capital by shifting into capital intensive export sectors whereas under autarky this possibility does not exist. Even when the existing technology would not enable sustained growth under autarky, **free trade makes long-term sustainable growth possible.**[68] In this respect trade can thus be regarded as an important determinant of growth.

Short-run economic growth versus long-run economic growth

As depicted above, economic growth theory distinguishes between **impacts on short-run economic growth**, which are **referred to as static contributions to growth**, and **impacts on long-run economic growth** which **change the steady-state growth path and which are dynamic in nature** and thus **refer to technological progress.**[69] In this respect it is crucial to note that not only increases in the steady-state growth rate are desirable, level effects shifting income on a higher level can be regarded as a major increase in welfare as well.

Referring to the studies analyzing the efficiency of public capital investment, their contribution to economic growth in

64 Aghion, P.; Howitt, P.; (1999), p. 367 ff.
65 Barro, R. J.; Sala-i-Martin, X.; (1995)
66 Pinna, A.; (1993), p. 5
67 Markusen, J. R.; (1995), p. 112
68 Markusen, J. R.; (1995)
69 Barro, R. J.; Sala-i-Martin, X.; (1995)

The efficiency of public capital investment

the neoclassical setting depends "on whether the marginal product of public capital exceeds, or, respectively, is exceeded by the marginal product of private capital."[70]

Concerning technological progress, it should be mentioned that it does not only refer to product innovation but also technological changes in production processes, which can lead to substantial efficiency gains as well. Changes in production processes, for example the introduction of just-in-time production, enabled by changes in transport conditions, are most likely to occur in the presence of an efficient transportation infrastructure guaranteeing a high degree of reliability.

In the subsequent part of the literature review the underlying mechanisms behind the relation between transport and economic growth are described in more detail. According to the above classification they are always divided into static and dynamic gains due to their differing nature.

2.1.3.2 *Gains from Trade*

Efficiency gains through trade

The possibility to trade has been an important driving factor behind the efficiency enhancing modification of production processes and has subsequently lead to significant gains in productivity.[71] Trade theory offers various theoretical concepts describing these modifications in production processes. Trade induced efficiency enhancing modifications of the production processes lead to changes in the production structure and the competitive environment. So changes in agglomeration patterns, production site sizes, location patterns and subsequently changes in the competitive environment are predominantly influenced by the possibility to trade and its conditions, like for example trade barriers, trade costs or transport costs.

Bernard et al. (2003) describe the following channels leading to higher productivity through falling trade costs:

"First, lower trade costs increase the probability of plant death, especially for lower productivity, non-exporting plants.

70 Aschauer, D. A., (1997a), p. 3

71 I am exclusively referring to the productivity functions that do not use infrastructure investment as a factor input but that model infrastructure investment equal to technical progress thus raising the whole function to a higher level.

Second, surviving high productivity, non-exporters are more likely to enter the export market, thus expanding their sales. Third, existing exporters, already the largest and most productive establishments, see their exports grow more quickly as trade costs fall."[72]

Concerning the classification into static and dynamic gains from trade, it can also be distinguished between "internal, i.e. caused by a rise in trade volumes and placing the economies at higher utility levels", which in our context refer to static gains, and "external gains e.g. by enhancing innovation and growth by increased levels of trade and transport innovation"[73] referring to dynamic gains from trade.

Static Gains from Trade

The gains-from-trade-theorem outlines the static gains from trade. They are referred to as static due to the fact that the production possibility frontier is not moved outward, there is only a movement *along* the curve. In order to reach dynamic gains from trade, more than a one-time movement along the curve is necessary, the curve has to move outward.

The production possibility frontier is not moved outward, there is a movement along the curve

The gains-from-trade theorem states that

"the value of free trade consumption evaluated at free trade prices exceeds the value of autarky consumption evaluated at free trade prices."[74]

Any price difference between the free trade prices and the autarky prices enables countries to realize gains from trade by
"selling what is relatively more valuable on world markets and buying on world markets what is relatively more costly to produce at home."[75]

Price difference between free trade and autarky prices

The gains from trade can be subdivided into **two categories** according to their sources: **on the one hand**, the gains from exchange resulting from countries' **differences in en-**

72 Bernard, A. B. et al.; (2003), p. 19
73 Steininger, K. W.; (2001), p. 35
74 Markusen, J. R.; (1995), p. 65
75 Markusen, J. R.; (1995), p. 73

dowments or preferences, and **on the other hand** the **gains from specialization** caused by the possibility to produce the goods where the country **has an advantage in manufacturing opposed to other countries**.

Trade induced by differences in endowments is largely referred to in the context of the **Heckscher-Ohlin model**. The **Ricardian trade model** describes trade as induced by comparative advantage. We speak of **comparative advantage** in producing a good X if its opportunity cost of producing good X in terms of good Y is less than in the other country. The gains from comparative advantage are reduced when transport costs increase, which results in a disincentive for multinational activity.[76]

Gains from Specialization

The degree of specialization is directly linked to the volume of trade

Bougheas et al. (1997) state that the degree of specialization is directly linked to the volume of trade:

"we focus on one unexplored aspect of the relationship between increased specialization and the stock of infrastructure, namely the link between the latter and the volume of trade."[77]

Romer's (1987) endogenous growth framework completely relies on specialization in production.

Horizontal versus vertical specialization

It can be distinguished between horizontal and vertical specialization. Horizontal specialization refers to trade in similar goods while in the course of vertical specialization firms split up their value chains, performing different stages of production in different countries. As vertical specialization increasingly gains in importance, I will give a more detailed description of vertical specialization in the following paragraph. Apart from this, vertical specialization tends to be hit by trade barriers especially hard due to the fact that one good is usually subject to various border crossings until it becomes a final product to be sold to consumers. Furthermore, gains from trade in the case of vertical specialization tend to be higher than from horizontal specialization.

76 Krugman, P.; Obstfeld, M.; (2001)
77 Bougheas, S. et al.; (1997), p. 170

Vertical Specialization

According to Ishii and Yi (1997) vertical specialization represents an important factor by which lower trade barriers and transportation costs explain the growth of world trade. Vertical specialization splits up the production process into several stages, so that a good crosses various borders before it reaches its final destination. As for horizontal specialization, the underlying causes are either comparative advantage or increasing returns to scale.

Vertical specialization splits up the production process

The effects of vertical specialization may lead to a non-linear surge in trade. For example, the first stage of production may be performed at home then the good is exported to another country and re-imported as a final good. During this process of exporting and re-importing border crossings and miles traveled are doubled, consequently the impact of tariffs and transport costs are much higher in the case of vertical specialization than for horizontal specialization.

The impact of trade barriers is much higher in the case of vertical specialization

Vertical specialization may provide an explanation for the growth in world trade without the usual assumptions of unrealistically high elasticities of substitution between home and foreign goods. Due to the fact that smaller elasticities lead to higher welfare gains from trade, gains from trade may be even higher than traditional estimates using high elasticities between home and foreign goods.

Apart from this, due to the layered production gains from trade tend to be higher than for horizontal specialization. This means that the magnitude of the gains resulting from trade differs according to its underlying causes, differing for vertical specialization as opposed to horizontal specialization. But despite the fact that gains from vertical specialization tend to be higher one needs to be cautious due to the fact that as the same good crosses one border twice, gains may be overestimated due to the possibility of double-counting. Empirical evidence shows that vertical specialization is also very important for developed countries.

Increasing Returns to Scale

Increasing returns to scale refers to the process during which increasing size, up to a certain degree, be it at firm, industry or industry group level, leads to increasing marginal returns by means of decreasing average costs with increases in plant or industry size. The occurrence of scale economies is independent of endowment differences of countries hence it is an important extension of the explanation of trade attributed to endowment differences.

Internal economies of scale

Internal Economies of Scale: If increasing plant size leads to increasing marginal returns at the firm level we speak of internal increasing returns to scale. Regarding the gains from trade, internal scale economies play an important role as their occurrence depends on plant size and trade induces the possibility to enable large-scale production.

External economies of scale

External Economies of Scale: Opposed to internal economies of scale, which arise at the firm level, external economies of scale arise at industry level or at the level of industry groups. They can occur even if the individual firms are producing under constant returns to scale. External economies of scale can be attributed to sector size which leads to increasing marginal returns although individual firms produce under constant returns to scale.

Krugman describes the **relationship between increasing returns to scale and transport costs** as follows:

"In a world characterized both by increasing returns and by transportation costs, there will obviously be an incentive to concentrate production of a good near its largest market, even if there is some demand for the good elsewhere. The reason is simply that by concentrating production in one place, one can realize the scale economies, while by locating near the larger market one minimizes transport costs. This point ... is the basis for the common argument that countries will tend to export those kinds of products for which they have relatively large domestic demand."[78]

The trade-off between transport costs and increasing returns to scale determines optimal plant size

This argument is usually referred to as the **home market effect** whereas it is crucial to add that it only holds true in a world of increasing returns to scale, under the assumption of diminishing returns to scale and strong domestic demand the good will be imported. The argument puts strong emphasis on the relationship between the gains from increasing returns to scale and the losses from transport costs. Decisions concerning optimal plant size and location are strongly influenced by the trade off between transport costs and increasing returns to scale. As long as the gains from the exploitation of increasing returns to scale supersede transport costs it is

78 Krugman, (1980), p. 955

beneficial to increase plant size and to concentrate production in one place.

Dynamic Gains from Trade

Speaking of dynamic gains it is referred to gains attributed to technological progress.

Markusen et al. (1995) describe how trade facilitates technological progress. A world of two economies is assumed, home and foreign, with a single final-goods sector in order to neglect considerations of comparative advantage. This allows us to concentrate on the scale effects of international trade.

Trade facilitates technological progress

Further assumptions are:

- The two economies are identical, whereas the ranges of intermediate inputs produced are entirely different.
- Each economy produces in isolation under conditions of autarky.
- The two economies are allowed to engage in intra-industry trade in differentiated intermediate inputs.
- There are no international flows of ideas.

In this setting, international trade does not alter the incentive to engage in research, and subsequently has no impact on the economy's rate of growth. This is due to the fact that that the number of varieties of final goods in each economy is the same as prior to trade. But there is a level effect as the number of varieties of the intermediate inputs is increasing and thus raises the product of labor.

If the setting is altered by allowing for international flows of ideas, the productivity of labor in research doubles. **Two effects altering the economy's equilibrium rate of growth** in a positive way can be distinguished: **a direct effect from the increase in the productivity of research and an indirect effect through the increased incentive to engage in research.**

Dynamic Comparative Advantage

Technology differences represent one possible source of comparative advantage. Traditional trade models assume technology as an exogenous variable, whereas in Schumpeterian models of growth it is included as an endogenous variable. Thus, the distribution of comparative advantage across

Technology differences as source of comparative advantage

nations itself becomes endogenously determined, depending on the degree to which knowledge spillovers occur nationally or internationally. If it is allowed for international knowledge spillovers,

"comparative advantage and the equilibrium allocation of resources between research, intermediate-input production, and low technology-goods production is uniquely determined by factor endowments. Therefore the standard Heckscher-Ohlin result carries over into a dynamic Schumpeterian model where technology is endogenous."[79]

Comparative advantage is partly determined endogenously depending on the degree of knowledge spillovers

Thus, in an endogenous model of economic growth even comparative advantage is partly determined endogenously depending on the degree of knowledge spillovers. As the formation of comparative advantage due to technological change is a dynamic process, it is often referred to as "dynamic comparative advantage."

Gains from Stronger Competition

Two positive effects of stronger competition

Two positive effects of stronger competition can be distinguished: an **increase in aggregate industry productivity as a result of falling trade costs and an increase in the incentive to innovate** due to foreign firms entering the market.

Bernard et al. (2003) describe the evolution of industry productivity in response to dynamic effects of a reallocation of activity across firms in response to changes in trade costs. They use several new firm-level models of international trade and find evidence of the positive effect of falling trade costs on industry productivity and show that

"as trade costs fall, industry productivity rises due to a reallocation of activity across firms: lower trade costs cause low productivity non-exporting firms to exit and high productivity non-exporters to increase their sales through exports, thereby increasing their weight in industry productivity."[80]

Market share is reallocated to more efficient firms therefore raising industry productivity

This holds especially true for trade in varieties which accounts for the major part of trade between developed countries. As the effect of lower trade costs is examined in a more direct way by linking them to individual firms within industries,

79 Markusen, J. R.; (1995), p. 382
80 Bernard, A. B. et al.; (2003), p. 2

direct and therefore more plausible evidence on the positive effect of lower trade costs on productivity and GDP growth is provided. Generally it can be concluded that market share is reallocated from inefficient to more efficient firms thereby raising industry productivity.

2.1.3.3 Empirical Evidence on the Relationship between Trade Volume and Income

Basically, it can be distinguished between studies analyzing the effect of trade on income respectively income growth and studies examining the effect of trade on productivity.

Trade and Productivity

Direct evidence on how trade affects productivity

Studies on the effects of trade on productivity claim to provide direct evidence on the extent to which increased trade may affect productivity and therefore GDP growth.

Islam (1998) states "that export industries are more susceptible to productivity improvements and these lead to more investment, higher profits and more rapid economic growth".

Bernard et al. (2003) analyze the effect of falling trade costs on aggregate industry productivity focusing on the microeconomic link between firm exporting and firm productivity. The underlying theoretical background stems from previous studies by Bernard et al. (2000), Melitz (2002) and Yeaple (2002). They introduce an increase in aggregate industry productivity, which is due to falling trade costs, as a key feature of three heterogenous firm, general equilibrium trade models. **Predictions are that lower trade costs cause low productivity non-exporting firms to exit and high productivity non-exporters to increase their sales through exports.** Bernard et al. test this hypothesis by linking trade costs by industry to plant-level data on the entire US manufacturing sector over a twenty year period. The data provides broad evidence for the fact that

"the aggregate industry productivity response to falling trade costs reflects the reallocation of activity across firms, away from low productivity non-exporters towards high-productivity exporters."[81]

81 Bernard et al.; (2003), p. 19

Alcalá and Ciccone (2003) conduct a macroeconomic study on the relationship between trade and productivity. The authors account for the endogeneity of trade by using instrumental variables. They follow Frankel and Romer (1999) and use a two-stage least-squares estimation. Firstly, they estimate

"a gravity equation for bilateral trade shares that uses countries' geographic characteristics and size only as explanatory variables. The second step of the approach aggregates bilateral trade shares predicted by the gravity equation to obtain a geography-based instrument for trade."[82]

Their findings show a causal effect of trade on productivity across countries, which is statistically and economically significant as well as robust.

Trade and Income

One prevalent methodology to provide evidence that increased trade is important for GDP (growth) is the estimation of cross-country growth regressions. Two-to-three-decade averages of GDP (growth) or GDP per capita (growth) are regressed on a number of variables including measures of openness or exports. Most studies produce positive and significant coefficients for the trade variables regressed on GDP.[83]

The studies of Frankel and Romer (1999) and Irwin and Trevio (2000), for example, produce positive and significant correlations between trade and income. Frankel and Romer (1999) examine the correlation between trade and income and claim that
"trade has a quantitatively large, significant, and robust positive effect on income."

Simultaneity problem between income and trade

They address the simultaneity problem between income and trade, which means that the relationship between income and trade might possibly not be mono-causal due to the endogeneity of trade. They overcome simultaneity bias by constructing an instrument that incorporates a country's geographic attributes correlated with trade but uncorrelated with income. Those measures are used to obtain instrumental variables estimates of the effect of trade on income. Re-

82 Alcalá, F.; Ciccone, A.; (2003), p. 8
83 Wälde, K.; Wood, C.; (2004)

sults provide evidence for the fact that OLS estimates do not overstate the effect of trade on income.

Concerning the causes of the positive effect of trade on income, Frankel and Romer (1999) found that 69 per cent of the positive impact of trade on income is induced by enhanced productivity, while only 31 per cent is to be attributed to capital accumulation.

Irwin and Trevio analyze

"the difference between OLS and two-stage least squares estimation of trade on income using data from several periods. ... The results reinforce the finding that trade serves to increase a country's real income and that OLS estimates understate the true effect of trade on income."[84]

Trade increases a country's real income

Rodriguez and Rodrik (2001) criticize Frankel and Romer's findings due to spurious correlation arising from the fact that the instrument used is possibly related with other variables that influence income through non-trade channels. They show that by re-running Frankel and Romer's income regression and introducing any of the summary indicators as a control, the positive impact of trade on income vanishes.

Noguer and Siscart (2005) re-run Rodriguez and Rodrik's regression

"using our instrument. Our trade estimate retains statistical and economic significance in all specifications. We conclude that the insignificance of the trade estimate in Rodriguez and Rodrik (2001) stemmed from the use of a weaker instrument. We also observe that the magnitude of the trade coefficient decreases substantially, which is consistent with their argument that failing to control for direct effects of geography imparts an upward bias on the trade estimate."[85]

Their findings show that **a 1 per cent increase in the trade share of GDP induces a 1 per cent increase in income** per **capita**. This result tends to be robust to the inclusion of control various variables but it is lower than the estimates of the previously cited studies.

Irwin and Trevio (2000) explore the statistical relationship between trade and income over three periods: the pre-World War I era, the interwar period, and the post-war era. They use a gravity specification that includes country dum-

84 Irwin, D. A. ; Trevio, M. ; (2000), p. 2

85 Noguer, M.; Siscart, M.; (2005), p. 452

mies which are assumed to capture GDP effects. They tested the instrument for exogeneity and found that the null of exogeneity could not be rejected. So they controlled for direct effects of geography in order to be able to exclude an upward bias on the trade coefficient. Estimating the regression with the inclusion of the direct geographic variables they found that trade has a statistically significant and economically important effect on income.

Concerning cross-sectional analysis, the problem arises that most studies take into account only

"a small number of explanatory variables in attempting to establish a statistically significant relationship between growth and a particular variable of interest. Given that over 50 variables have been found to be significantly correlated with growth in at least one regression, one cannot rely on the findings of any one study."[86]

Omitted variable bias is likely to be more severe in a cross-sectional analysis because the omitted variables tend to change more across countries than across time. Furthermore, if multiple errors vary more across countries than across time cross-section estimates are more likely to be biased than time-series estimates.[87]

Apart from this, regarding the effects of trade across countries, recent empirical work has tended to assume the same effect across countries, although in practice there may be quite substantial differences according to different specialization patterns of the countries.

2.1.3.4 Conclusions

Static and dynamic gains from trade

The possibility to trade leads to substantial welfare gains. They can be subdivided into static and dynamic gains from trade depending on whether the production possibility frontier is moved outward by technological progress, which then refers to dynamic gains from trade, or whether a movement along the production possibility curve is described, referring to static gains from trade.

Empirical evidence shows that more open economies reach a higher per capita income than less open countries.

86 Pinna, A.; (1993), p. 4
87 Ibrahim, I.; (2002)

"The policy implications of these studies seem to be clear. As trade, ..., continues to have a statistically significant and positive impact on growth any measures that promote trade seem advisable."[88]

Trade has a positive and significant impact on growth

None of the above depicted mechanisms leading to welfare gains from trade can be separated from each other. They are closely intertwined in the sense that one factor is conducive to the development of another factor and vice versa. Specialization for example is an important precondition for exploiting increasing returns to scale, vice versa increasing returns to scale lead to further specialization of production.

Due to the gains from trade described above, the conclusion can be drawn that the volume of trade between countries is a major determinant of output.

In chapter 2.2.2 the relationship between trade volume and output (growth) is examined empirically. From trade (export tables) and output data on regional level, the effect of a decrease in trade volume due to changes in transport costs is studied. The hypothesis is that increases in transport costs cause a decrease in trade volume, which again is assumed to trigger a decrease in aggregate output.

2.1.4 Location decision, Production Structure and Economic Growth

Infrastructural aspects strongly influence location decisions. Accessibility is a necessary precondition for business activity. Demand as well as supply linkages depend on the possibility to transport goods and resources.

"On its own, transport infrastructure is a second order location variable where there is a well-developed network, but in conjunction with other factors it may 'tip the balance' in favor of the (marginally) more accessible location."[89]

The importance of transport for location decision seems to be an obvious fact. For the present analysis the question arises how location decisions influence overall economic growth.

Depending on the focus of the analysis the importance of location decisions for economic growth can be determined.

The importance of location decisions for economic growth

88 Wälde, K.; Wood, C.; (2004), p. 9

89 Banister, D.; Berechman, Y.; (2000), p. 331

If attention is mainly directed towards regional development, firms' locating in the region under scrutiny is certainly a crucial aspect. This also holds true if the focus is a national one, then competition between locations arises on international grounds between nations. Hence, for the development of a certain area the attractiveness of several locations compared to each other represents the determining factor.

Concerning overall economic growth, the productivity gains made possible by lowering the constraint distance puts on production decisions are relevant. As decisions of location and plant size are taken by weighing costs and benefits, the higher the constraint represented by the costs of transporting goods and resources the greater the impact on decisions. This means that possible efficiency gains, for example from large-scale production, can not be fully exploited due to transportation costs. As mentioned above, the resulting trade-off between transport costs and gains form concentration strongly influences production decisions.

"The microeconomics of location depends on two interrelated features, the existence of increasing returns as the raison d'être of a spatial economy and the differential importance of transport costs to different sectors. Increasing returns justify the spatial separation of production and hence the localization of industry. This implies the definition of minimum efficient technical scales of production. ... firms must be able to maintain a minimum size market area."[90]

Three basic theoretical approaches in location theory

In location theory three basic theoretical approaches can be distinguished. The Neo-Classical Theory (NCT) is based on assumptions of perfect competition, homogenous products and non-increasing returns to scale. Location is exclusively determined by the spatial distribution of natural endowments which are exogenously determined. Because of the assumption of zero trade costs, location is independent of the spatial distribution of demand. If this assumption is abandoned, the spatial pattern of economic activity is correlated with the level of trade costs.

On the other hand, New Trade Theory (NTT) only accepts one exogenous factor which is market size. Spatial dispersion of activity is determined by the degree of imperfect competition, differentiated products and increasing returns. Hence, as market size is the only exogenous variable, its importance for the spatial distribution of activity is significant. As market size is significantly affected by transport costs, this approach

90 Vickerman, R.; (1995), p. 226

might be suitable for an analysis of the effect of transport costs on the spatial pattern of economic activity.

In the third approach, the New Economic Geography (NEG), spatial patterns of activity are completely determined endogenously. This approach is characterized by multiple equilibria and outcomes.[91]

Two basic sources of gains

Summing up, concerning decisions on plant location and size, two basic sources of gains leading to economic growth can be distinguished. **Gains from agglomeration** which arise **due to firms locating close to each other** and **gains from concentration**, which refer to the gains made possible by being able to choose **the optimal production plant size**.

2.1.4.1 Gains from Agglomeration

Historical data provide much evidence on a positive correlation between agglomeration and growth of economic activities, which suggests a positive relationship between these two variables.[92]

In his early work Marshall (1920) distinguishes between three basic sources of gains from agglomeration:

- *"localization provides a pooled market for workers with specialized skills*
- *facilitates the development of specialized inputs and services*
- *enables firms to benefit from technological spillovers".*[93]

Static gains

Larger market size

Producers of intermediate goods benefit from a lager market size due to the close proximity of other firms using their goods as inputs.

Dynamic gains

Knowledge spillovers and incentives to innovate

1. Knowledge spillovers are more likely to take place between firms that are located closely to each other, independently of the exchange being intra- or inter-industry.

91 Brühlhart, M.; (1998)
92 CEP II, document de travail no. 96–14
93 Head, K. et al.; (1995), p. 224

2. Stronger competition induces incentives to innovate quickly which in turn promote technological progress and hence long-run economic growth.[94]

Empirical work shows that agglomeration is especially important for innovative activity as it tends to cluster where production activities concentrate and is even more spatially concentrated than production itself.

CEP II construct a model that introduces growth and geographic agglomeration as a self-reinforcing processes in the form of circular causation, thus, "growth brings spatial agglomeration which itself fosters growth."[95]

2.1.4.2 Gains from Concentration

Gains from concentration refer to increasing returns to scale and learning curve effects and are described in more detail in the part on gains from trade, chapter 2.1.3.2.

2.1.4.3 Plant Location, Plant Size and Transport Costs

Concentration and de-concentration of activities

Various models of location theory tell us that firms have a propensity to locate in places where economic activities have already been established. Strong competition, high transport costs and land rents lead to de-concentration of activities whereas low transport costs, high economies of scale and a high degree of product differentiation leads to concentration of economic activities. Thus, the development of agglomeration is the result of an interaction of demand and supply specific factors.

"Consequently, the space-economy has to be understood as the outcome of the interplay between centripetal and centrifugal forces within a general equilibrium framework accounting explicitly for market failures."[96]

The magnitude of transport costs has ambiguous effects

According to Krugman and Venables (1990), the magnitude of transport costs can have ambiguous effects. In their model, high transport costs lead to dispersed production, the elimination of transport costs may induce production being

94 Gao, T.; (2003)

95 CEP II, document de travail no. 96–14,

96 Ottaviano, G. I. P.; Thisse, J.-F.; (2003)

concentrated in low production cost and thus peripheral locations. This is due to economies of scale, which can better be exploited in an economy with low transportation costs and locations with specialization tendencies may benefit more from a reduction in transport costs than other locations. A partial reduction of transport costs could increase concentration in core locations as long as the greater scale economies outweigh transport costs. Thus, low transport costs affect location decision in an opposite direction as agglomeration economies.

However, agglomeration economies may as well be promoted by low transport costs, in the case where production is organized in one location with suppliers located in close proximity. So, low transport costs enable firms to produce centrally. But due to the suppliers locating close to the production facilities, transport distances in production may be lowered. **Two trends can be distinguished: for distribution, transport intensities will be rather high, while for production, transport intensities will be low due to agglomeration economies.**

According to Krugman and Venables (1995) sectors with a high initial level of transport costs become more concentrated as transport costs decrease whereas the opposite is true for sectors with low initial transport costs. Their argument works as follows: At prohibitively high transport costs, goods are non-tradeable, hence producers locate close to consumers, activity is de-concentrated. Due to falling transport costs, goods become tradeable, thus production is concentrated. But the gains from agglomeration economies may be outweighed by the losses due to congestion. There is a trade-off between congestion costs and gains from agglomeration.[97]

So, concerning **agglomeration economies, two different tendencies** can be distinguished. On the one hand, **agglomeration factors lead to proximity**, whereas **in an economy with a well-connected infrastructure diseconomies of agglomeration may operate**. With low transportation costs many services can be provided remotely and there is a greater flexibility in the location decision of firms.

The greater this flexibility the lower is the influence of geographic characteristics on production decisions. Subsequently, the importance of production specific aspects (for example the exploitation of economies of scale) in decisions increases.

97 Desmet, K.; Fafchamps, M.; (2003)

According to Desmet and Fafchamps (2003, p. 7)

"The literature has pointed to the drop in transport costs as one of the main forces affecting the nature of agglomeration economies."

2.1.4.4 Conclusions

Concerning the impact of transport related factors on plant location and size, two different aspects can be distinguished:

1. **Low transport costs lead to concentration of production in one place.** They foster large plant sizes and the possibility to reap the benefits from scale economies. Decisions on plant size and location are less dependent on the magnitude of the distance to consumers the lower transport costs are. But not only distance to consumers becomes less relevant, also upward linkages to suppliers are no longer regarded in dependence to distance. So **low transport costs work against economies of agglomeration**.
2. Concerning agglomeration economies good transport infrastructure may also have a positive impact as it opens up new possibilities of industrial organization. For example just-in-time production which strongly depends on an efficient transport system favors agglomeration economies as firms' locating close to each other reduces the risk of production breakdown. Apart from this, the effect of transport costs on agglomeration is not essentially a negative one, as agglomeration economies increase distance to consumers. Hence, **low transport costs might also be conducive to agglomeration economies in the case where longer distances to consumers are mitigated by transport cost reductions**.

The relationship between the degree of concentration and trade volume

The first aspect can by examined in an analysis of trade flows. It can be assumed that the higher the degree of concentration of production the higher will trade volume be as distance from production site to consumers increases.[98]

The second aspect can not be incorporated in an analysis of trade volume as this might actually decrease due to ag-

98 What might pose a substantial difficulty in this respect, is national concentration of production which cannot be examined in an analysis of international trade flows.

glomeration economies. In this respect the above described analysis of trade is no longer sufficient for the analysis of the impact of transport on economic development. But due to the relatively restricted focus and resources, **the present study focuses exclusively on gains from improved transport infrastructure that emerge from increasing trade volumes.** As other gains are not scrutinized in detail in the present analysis, results might underestimate the positive effects on income and income growth respectively.

The present study analyzes gains from transport through trade volumes

2.2 Empirical Evidence on the Relationship between Transport, Trade and Economic Growth

Previous research on the relationship between transport infrastructure investment and economic growth, measured as aggregate output growth, produces widely differing results due to differences in the aggregation level of data, in data sources and in model specifications. Although in most cases there is a positive and significant impact of transport infrastructure investment on output growth, several studies conclude insignificant positive or even significant negative impacts of transport infrastructure investment on output growth.

Previous studies focus on infrastructure investment and output growth

Besides the wide differences in results obvious specification problems question their plausibility. The problem of reverse causation is especially worrying because if the direction of causality is not clear, the conclusion of positive impacts of transport infrastructure investment on output growth might be completely wrong as output growth causes increased investment instead of vice versa.

The fact that most studies produce positive and significant impacts of infrastructure investment on aggregate output growth does not necessarily lead to the conclusion that further infrastructure investment is efficient. Under the assumption that production factors are not working to full capacity investment ALWAYS leads to output growth no matter whether people are employed to dig wholes into the ground or to build roads. As soon as wages are paid for the work that is done, GDP increases independently of its utility for other branches of the economy or for individuals. What really matters concerning the positive impact of infrastructure on GDP is actually perceived infrastructure quality by the users which results in transport costs, time, reliability etc.

The underlying mechanisms are not analyzed

Summing up, models that introduce transport infrastructure investment as independent variable and output as dependent variable, simply measure the magnitude of the impact but not the underlying mechanisms leading to this impact.

Now, I could endeavor on producing another macroeconomic study analyzing the impact of transport infrastructure investment on output trying to handle the specification problems but, especially in the case in which results are quite controversial, it seems to be more fruitful to examine the underlying mechanisms in order to be able to better estimate impacts and the causal directions of these impacts.

Elaborating on hypothesis for research one could argue that trade is directly influenced by the possibilities and conditions to transport goods. This impact has been examined by several studies, most often it is compared to the influence of tariffs. Results reflect a consensus on the fact that the magnitude of the impact of transport costs can be compared to the constraint imposed on trade by tariffs.[99] The latter have been subject of analysis of a vast amount of research, as regards to the former further research is needed.

The present analysis is separated into two parts

In the subsequent analysis, the impact of changes in transport costs on trade volume and the influence of changes in trade volume on output are examined. This analysis is separated into two parts: The impact of transport costs on trade volume (chapter 2.2.1) and the impact of changes in trade volume on output (chapter 2.2.2).

2.2.1 Transport Costs and Trade Volume

Empirical studies are dominated by gravity models

The empirical analysis of international trade flows is dominated by the use of gravity models[100] its outstanding empirical success being widely confirmed in the literature.[101] Chen (2004) considers it to be "the most robust empirical relationship known in explaining the variation of bilateral trade flows".

The underlying theoretical framework was first developed by Anderson (1979). His model produces the basic frame-

99 Rossi-Hansberg, E.; (2003);

100 The first authors applying a gravity equation empirically were Tinbergen (1962) and Pöyhönen (1963).

101 Deardorff (1984); Evenett, S. J. and Keller, W.; (1998); Leamer, E.; and Levinsohn, J.; (1997); Stijns, J. P.; (2003)

work of the gravity equation, relating the volume of trade to the income of two countries and a distance variable between these countries:

$$X_k^{ij} = \alpha_k Y^i Y^j d^{ij} U_k^{ij} \qquad (2.6)$$

X_k^{ij}... denotes exports of good k from country *i* to country *j*
$Y^i; Y^j$ ... denote country *i*'s respectively country *j*'s GDPs
d^{ij}... denotes distance between countries i and j
U_k^{ij} ... denotes a lognormally distributed error term

Anderson (1979) derives the gravity equation from the properties of expenditure systems. Identical homothetic preferences are assumed, which means that preferences neither vary between countries nor with different levels of income. This assumption has been proved to be rather robust. Apart from this, the model relies on complete specialization hence each good is produced in only one country.

Analyzing the causes of specialization

More recent research employs the gravity equation to test different theories of trade analyzing the causes of specialization. They pose the question whether trade emerges due to factor endowment differences in a framework of constant returns to scale or due to product differentiation in a framework of monopolistic competition with increasing returns to scale. The gravity equation can be derived from both theories, yet most studies rely on the framework of monopolistic competition with increasing returns.

"The product differentiation model states that countries trade with each other even if the varieties of a good are substitutable because consumers prefer an increasing number of choices, under an assumption that each firm produces a variety of a good (monopolistic competition) with an increasing return to scale technology."[102]

The framework of monopolistic competition

The basic framework of monopolistic competition with product variety was analytically developed by Dixit and Stiglitz (1977).[103] Firms produce distinct, imperfectly substitutable goods, so product variety is the critical component of competition. Due to the fact that goods are imperfect substitutes

102 Kim, M. K. et al.; (2003), p. 1
103 Although several authors developed similar structures before, the major breakthrough was reached by Dixit and Stiglitz. (1977)

each firm has some degree of monopoly power which allows them to set their prices above marginal costs. The average cost function is downward sloping and so indicative of scale economies. The possibility to reach broader markets allows firms to move down their average cost curves. In the short run they will earn higher profits, in the long run more firms will enter the market, which shifts downward the demand curves of the existing firms. In the new equilibrium the number of firms and thus variety is larger than before. Of course this theory only works under the assumption that variety per se increases utility, which seems to be true especially for developed countries with an ever increasing number of varieties of different products.[104]

While for the study of trade volumes the theoretical foundation of the gravity equation is not of utmost importance for the estimation results in the first part of the analysis; for the second part of the present study, in which the welfare impacts of changes in trade volumes are scrutinized the specification of the theoretical framework matters.

2.2.1.1 Theoretical Considerations – The Gravity Equation in a Framework of Monopolistic Competition[105]

Various trade theories analyzing trade flows

As already depicted above, it can be distinguished between various trade theories analyzing international trade flows. Depending on the time when they were developed, they are based on different assumptions concerning competition and the nature of products. Concerning the impact of transport costs on welfare, different trade theories reach different conclusions. Only new trade theories will be regarded in this context due to the fact that the others rely on assumptions that are too unrealistic for the developed countries of today. Both imperfect competition and product differentiation are taken into account. Due to the already high degree of complexity, transport is not modeled as a separate sector, transport costs are modeled according to the iceberg model.

Neo-Heckscher-Ohlin theories continue to apply the assumption of perfect competition despite they introduce product differentiation by quality. When transport costs are introduced, the range of quality levels is reduced.

104 Lakshmanan, T. R.; Anderson, W. P.; (2002)
105 For the derivation of the theoretical model cf. Feenstra, R. C.; (2002)

Other approaches introduce both imperfect competition and product differentiation simultaneously. The model of monopolistic competition is by far the most prevalent in new trade theory due to its good applicability but also due to its assumptions appearing to be relatively close to reality. Internal economies of scale at firm level and the assumption that increases in variety per se increase utility, lead to a monopolistic market structure. In this model, the introduction of transport costs leads to a decline in the number of varieties produced. Due to the simplifying assumption of fixed output per firm, the quantity of output of each variety is not affected in this model. **The model for the empirical part of this analysis will be based on the assumption of monopolistic competition with product differentiation.**

As the gravity equation can be derived from a monopolistic competition model with increasing returns to scale, intra-industry trade can be explained. Due to the fact that the present analysis concentrates on the EU 15 countries, intra-industry trade is likely to make up a major part of total trade.[106] When Feenstra et al. (2001) calibrate their model they find that the theoretical foundation is more solid for differentiated goods than for commodities, which is another confirmation for the applicability of the gravity equation for the analysis of trade between developed countries. For the part of trade not being in differentiated goods, which is assumed to be rather small for the EU 15 countries, the fact that the gravity equation is consistent with various theories of trade legitimates the use of the gravity equation also for this part of traded goods. Feenstra et al. (2001) have shown that the gravity equation even works for trade in homogenous goods.[107]

In the subsequent analysis a model of monopolistic competition with product differentiation and increasing returns to scale will be used. Complete specialization is assumed in the sense that each country exports varieties of the differentiated product to another country but each variety is only produced by one country due to the fact that firms can differentiate costless between varieties so that under the profit maximizing condition it is best for them to produce different varieties. Opposed to this, in autarky different countries would produce the same varieties. Both features complete specialization and intra-industry trade correspond to a Heck-

106 Evenett, S. J., and Keller, W.; (1998); Kim, M. K.; et al. (2003)
107 Stijns, J. P.; (2003); Kim, M. K. et al. ; (2003)

scher-Ohlin model with a continuum of goods. The assumption that there are many more goods than factors allows for complete specialization in different product varieties across countries. In this framework, trade patterns can be described by a traditional gravity equation.

As in the standard model developed by Anderson (1979) identical and homothetic preferences describe the demand side. In its basic form, the gravity model predicts bilateral trade volume between two countries to be directly proportional to the GDPs of the countries under the assumption of free trade and identical prices for both countries. The assumption of zero barriers to trade, hence neither transport costs nor tariffs, will be loosened later on. Under these basic assumptions bilateral trade volume can be derived as a proportion of the purchasing country's GDP.

In a multi-country framework $i,j = 1, ..., C$ denote countries and $k = 1, ..., N$ denotes products, each variety counting as a distinct product. y_k^i denotes country i's production of good k, which corresponds to the value of production because prices are normalized to unity as they are the same across all countries. Total GDP in each country is $Y^i = \sum_{k=1}^{N} y_k^i$, world GDP is $Y^w = \sum_{i=1}^{C} Y^i$. Under the assumption of balanced trade, s^j which corresponds to country j's share of world expenditure at the same time equals its share of world GDP, $s^j = Y^j / Y^w$, then exports from country i to country j of product k are given by $X_k^{ij} = s^j s_k^i$. Summing over all products k gives

$$X^{ij} = \sum_k X_k^{ij} = s^j \sum_k y_k^i = s^j Y^i = \frac{Y^j Y^i}{Y^w} = s^j s^i Y^w = X^{ji} \quad (2.7)$$

Introducing Trade Barriers into the Gravity Equation

Under trade costs the assumption of price equalization across countries no longer holds. In order to model different prices across countries, a specific utility function needs to be assumed. A possible way to do this is to assume a Constant Elasticity of Substitution (CES) specification. c_k^{ij} denotes exports of good k from country i to country j and at the same time total consumption of good k in country j due to the assumption of complete specialization. Country $i=1, ..., C$ produces N^j products. Elasticity of substitution between varieties is equal to $\sigma > 1$. Thus, utility for country j is

$$U^j = \sum_{i=1}^{C} \sum_{k=1}^{N^i} (c_k^{ij})^{(\sigma-1)/\sigma} \quad (2.8)$$

All products exported by country *i* sell for the same price p^{ij} in country *j*, usually this includes all trade costs on a CIF basis. Opposed to this, goods produced and sold in country *i* are FOB prices.

Since the present analysis is restricted to countries within the Single European Market and to the examination of changes in transport costs, transport costs are the only barriers to trade that will be introduced into the gravity equation. So instead of differences between CIF and FOB prices, real transport costs are used. Transport costs are modeled as Samuelson iceberg costs, which means that transport is not regarded as a separate sector, it is assumed that transport consumes a fraction of the good transported. Whereas $p^{ij} = T^{ij} p^i$, and $T^{ij} \geq 1$ so that ($T^{ij} - 1$) units melt along the way.

We assume that there are no differences in quality between varieties, so prices p^{ij} across varieties are equal, hence consumption too is equal over all products $k = 1, \ldots, N^i$ sold by country *i*, such that $c_k^{ij} = c^{ij}$.[108] The utility function can be specified as follows

$$U^j = \sum_{i=1}^{C} N^i (c^{ij})^{(\sigma-1)/\sigma} \tag{2.9}$$

Consumers maximize their utility subject to the budget constraint

$$Y^j = \sum_{i=1}^{C} N^i p^{ij} c^{ij} \tag{2.10}$$

Because balanced trade is assumed, Y^j, aggregate expenditure equals income in country *j*. If utility is maximized subject to the budget constraint demand for each product c^{ij} can be expressed as

$$c^{ij} = (p^{ij}/P^j)^{-\sigma}(Y^j/P^{j)} \tag{2.11}$$

Pj, country *j*'s overall price index is defined as

$$P^j = \left[\sum_{i=1}^{C} N^i (p^{ij})^{(1-\sigma)} \right]^{1/(1-\sigma)} \tag{2.12}$$

The total value of exports from country *i* to country *j* will be

$$X^{ij} \equiv N^i p^{ij} c^{ij} \tag{2.13}$$

108 This restriction might be eliminated for an analysis of differences in transport costs across sectors.

Considering price indexes this yields

$$X^{ij} = N^i Y^j \left(\frac{p^{ij}}{p^j} \right)^{1-\sigma} \tag{2.14}$$

In reality the number of products N^i is unobservable. But under the assumption of symmetry, firm output is fixed at $\bar{y}$, so the number of varieties is given by

$$N^j = \frac{Y^j}{\bar{y}}$$

By using the zero-profit condition we get

$$X^{ij} = \frac{Y^i Y^j}{p^{i\sigma} \bar{y}} \left(\frac{T^{ij}}{P^j} \right)^{1-\sigma} \qquad (2.15)\ (5.10)$$

Where $\bar{y}$ denotes firm output which is fixed (for details see Feenstra, 2002), country price indices, p^i and P^j, are measured with GDP deflators. Taking logs and first differences we get

$$\Delta \ln X^{ij} = \Delta \ln(Y^i Y^j) + (1-\sigma)\Delta \ln T^{ij} - \sigma \Delta \ln p^i + (\sigma - 1)\Delta \ln p^j \text{ [109]} \tag{2.16}$$

In the case of transport costs, there is a welfare loss caused by the fact that resources are used up in transport. This reduces the number of varieties produced. Due to the simplifying assumption of fixed output per firm, the quantity of output of each variety is not affected in this model.[110]

Concerning comparative advantage a country relatively well endowed with capital is more likely in setting up firms and therefore in running varieties than a less capital-abundant country. Under the assumption of Dixit–Stiglitz preferences, this implies that a country with a high range of different varieties is less affected by changes in transportation costs due to the assumption of love-for-varieties.[111]

For the estimation in the present study, neither GDP nor trade volume will be deflated due to the fact that inflation rates do not differ substantially between the analyzed countries within the relevant period of time. So the price indices, p^i and P^j, are no longer needed. This yields the following estimation in absolute values:

109 Feenstra, R. C.; (2002)
110 Steininger, K. W.; (2001)
111 Egger, P.; (2001)

$$\ln X^{ij} = \alpha + \beta_1 \ln Y^i + \beta_2 \ln Y^j - \gamma \ln T^{ij} + \varepsilon_{ij} \tag{2.17}$$

and in first differences:

$$\Delta \ln X^{ij} = \beta_1 \Delta \ln Y^j + \beta_2 \Delta \ln Y^i - \gamma \Delta \ln T^{ij} + \varepsilon_{ij} \tag{2.18}$$

2.2.1.2 *Empirical Estimation*

Building hypotheses based on theory

In the following part of the study, the impact of transport costs on trade volume is analyzed empirically. Building hypotheses based on theory, **bilateral trade volume is positively dependent on the exporter's and the importer's GDPs and negatively dependent on transport costs and other trade barriers.**

Methodology

In this empirical study a gravity equation as delineated above (2.17) is used. In order to analyze the impact of transport costs on trade volume – following theoretical propositions – export volume is regressed on both the exporter's and the importer's GDPs and the transport costs between the two countries. As trade flows within the European Union are analyzed transport costs are assumed to be the only significant trade barrier in this context.

The following equation will be estimated:

$$\log TV = \alpha + \beta_1 \log GDP_{\exp} + \beta_2 \log GDP_{imp} - \gamma \log TC + \rho\delta \tag{2.19}$$

TV denotes export volume, GDP_{exp} is the exporting country's *GDP*, GDP_{imp} the importing country's *GDP*, *TC* are transport costs and *6* is a year dummy variable. By performing a simple ordinary least squares regression, the coefficients β_1, β_2 and *6* show the impact of the exporting country's *GDP*, the importing country's *GDP* and of transport costs respectively on trade volume.

Data

Regional data for Germany and Italy

To examine the effects of transport activities on economic growth regional data for Germany and Italy were collected.

A significant part of North-South trade within the European Union takes place between Italy and Germany. An analysis at the regional level using transport costs as an indication of trade barriers allows being much more accurate than using aggregate national data. Export and import data is available for Germany as well as for Italy on a regional level, but comparable regional data for other EU countries were not available.

Regional trade data

Italian trade data was obtained from the official website of the Italian National Statistics institute ISTAT[112], German trade data stems from DESTATIS, the German National Statistics Institute.

These data only give aggregate exports (imports) from/to every German "Bundesland" to/from Italy and vice versa, for example exports from Bavaria to Italy or imports to Tuscany from Germany. On the basis of this data it is only possible to get a breakdown on a regional level for one country at a time. So these data need to be matched in order to get trade flows from/to every "Bundesland" to/from every region in Italy.

As export flows are introduced into the gravity equation it is thus necessary to get the export data on a regional basis. From Italian import data the percentages that go into every region in Italy from the overall imports from Germany can be derived. Therefore, the percentages that each Italian region absorbs from the overall exports from Germany, the exports from each German "Bundesland" only need to be multiplied by the percentage that the Italian region in question absorbs in order to get export trade data from every German "Bundesland" to every Italian region.

Regional GDP data

Regional GDP data from Italy was obtained from the Chamber of Commerce in Bolzano (Camera di Commercio di Bolzano), regional GDP data from Germany is available on the official website of the National Statistics Institute, DESTATIS.

A substantial amount of research on transport and trade focuses on sea transport. This is because a major part of the trade literature concentrates on trade between developed and developing countries or exclusively on developing countries. As the present study focuses on developed countries, I concentrate on land, in particular on road transport as this is one of the most important modes of transport for shipping goods within the European Union.[113]

112 www.istat.it

113 European Commission; (2003a)

The costs of land transportation are not linearly correlated to distance opposed to costs resulting from sea transport which tend to correspond to transportation distances more closely. Therefore, **it is more precise to take transport** costs **instead of distance as a proxy for transport costs** as is usually done in empirical studies estimating the impact of transport costs on trade volume. Land transportation costs are influenced by various other factors such as road quality or capacity in relation to usage.

Instead of using a distance measure as a proxy for transport costs I introduce observed transport cost data. Studies using distances often apply the "great circle" formula, which means that instead of using existing routes to measure distances, longitude and latitude of the capital or "economic centre" are used. This means that measures underestimate real distances goods need to travel.[114]

Transport cost data

In order to get more exact measures for transport costs, I do not use one single transport cost measure for each country. Rather, **each country is divided into four geographic regions.** In the present analysis, Germany and Italy are divided into four regions and transport costs for trading between two regions are proxied by the *actual* transport costs from one major city which is located in the centre of each region to another major city located in the centre of each region in the other country e. g. actual transport costs from Munich to Milan. The splitting-up of the two countries is done in order to get more exact results and to get more degrees of freedom in the regression analysis. This yields 32 trade flows for each year.

Transport cost data was obtained from an international forwarding agency located in Austria.[115] It has operations all over the European Union, a considerable amount of its operations are carried out between Italy and Germany. Due to its long and extensive trading experience in these two countries and also due to the large volume that is handled by this company, it can be assumed that this is quite reliable data. The transport cost data reflect the costs of shipping one payload, which is around 24 to 27 tons depending on the nature of the goods transported, from the centre of one region in Germany to the centre of another region in Italy. By region I do not refer to the German "Bundesländer" nor to the Italian provinces. Due to the fact that the areas of both the German "Bundesländer" as well as of the Italian provinces are rather

114 Head, K.; (2000)

115 Due to the company's request, its name will not be mentioned in the reference section.

small, differences in transport costs do not accrue to every Bundesland or to every province respectively. They only matter when distances are big enough to be a relevant factor in transport costs. So Germany as well as Italy has been divided into four regions, west, east, south and north. This yields 16 different routes between Germany and Italy with differing transport costs on each route.

The data cover three years

The above depicted data cover three years: 1993, 1999 and 2003.

Results from the Regression in Levels

In the first regression, the nominal GDP of the exporting region, the nominal GDP of the importing region and the nominal transport costs were regressed on the nominal export volume between two regions. Using real values would not make a difference as all variables are either in nominal or in real values. Logarithmic values are taken for all variables before running the regression.

The data for three years of observation, 1993, 1999 and 2003 respectively, is introduced in a row. As four regions in every country are examined, this yields 16 observations for trade flows in one direction for one year, thus, 48 observations for all three years for trade flows in one direction, 96 observations for all trade flows in both directions.

Results verify predictions from trade literature

Because of possible heteroscedasticity robust standard errors are provided. Table 2.1 shows that results verify the predictions of the trade literature. These bilateral trade data indicate that export volume is significantly related to the GDP of the importing country. The exporting country's GDP coefficient shows a 1.229 % increase in export volume to the examined importing country when the exporting country's GDP increases by 1 % . The importing country's GDP shows that export volume from the exporting to the importing country increases by 1.48 % when the importing country's GDP increases by 1 % .

What is more important for the present study is that **export volume significantly depends on transport costs**. The transport cost coefficient shows that trade volume is reduced by 2.06 % when transport costs rise by 1 % . All coefficients are highly significant. The model is statistically significant. With an R^2 of 0.8179, 81 % of the variability of the export volume can be explained with the variables used in the model. An inspection with regard to multicolinearity indicates no problems.

Table 2.1 Effects of changes in transport costs on log export volume

Constant	–24.026 (7.234)
log transport costs	–2.061 (0.182)
log GDP exporting region	1.229 (0.214)
log GDP importing region	1.479 (0.149)
Dummy 99	–0.415 (0.143)
Dummy 03	–0.785 (0.138)
R^2	0.8179
No. of Observations	96

Time-fixed Effects

In order to test for fixed effects over the years, time dummies were introduced. As the observations cover three years, two dummy variables are used. After the inclusion of the dummy variable the coefficients are still significant, their magnitude did not change significantly.

Time dummies were introduced

Results from the Regression in Differences

One basic assumption for the linear regression model is that time series are stationary. In practice, this means that the time series must not have stochastic trends.

If non-stationary time series are regressed on each other this usually results in spurious correlation. In this case the least squares regression often produces high R^2 values and significant t-statistics but these are actually artificial and misleading results which have no real meaning. A significant relationship is found where actually none exists. This problem frequently occurs with many macroeconomic variables, such as GDP, industrial production, employment, consumer prices, wages or common stock prices as these variables tend to follow trends. Granger and Newbold (1974) who initially coined the phrase "spurious regression" indicate that estimating the regression in first differences might eliminate this problem if both series are integrated of order one.[116]

Since testing for non-stationarity is not possible because the time series are much too short I pragmatically use first differencing for eliminating potential problems and also examining the robustness of the regression.

Thus, in order to test for the reliability of the regression in levels, a regression in differences is performed. This means that instead of using absolute values the differences of the logs of trade volume, of GDP and of transport costs between the years 1993, 1999 and 2003 are used. Logs of the variables as well as their first differences are taken. To allow for a time trend an intercept is included.

The regression in first differences also yields significant results. This strengthens the results of the previous section.

2.2.2 Trade Volume and GDP

Extensive previous research on the relation between trade volume and income

There is extensive research on the relationship between trade volume and income. Cross-country ordinary-least-squares growth regressions are a prevalent methodology used to analyze the impact of changes in export volume on income or income growth. Usually averages of GDP (growth) or GDP per capita (growth) over several decades are regressed on a number of variables including measures of openness or exports. These studies generally indicate that the coefficient on the trade variable is positive and significantly different from zero.[117]

Strong and positive effect of increases in trade volume on income

In a cross-country analysis Frankel and Romer (1999) for example estimate that a one percentage point increase in the trade share of GDP increases per capita income by 2 per cent. Their results underwent various tests concerning their reliability. Simultaneity bias was one of the major problems they were confronted with. They found an instrument, which I described in more detail in the literature review, to overcome simultaneity bias. Though they corrected for simultaneity bias, Frankel and Romer's findings were criticized on the basis of spurious correlation.

Noguer and Siscart (2003) rerun the regression with summary indicators as control variables. Results are still positive and significant but the coefficients are a little lower than

116 Griffiths, W. E. et al.; (1992)

117 Wälde, K.; Wood, C; (2004)

Table 2.2 Effects of changes in transport costs on log export volumes (estimation in first differences)

Constant	-0.087 (0.092)
Δlog transport costs	-1.744 (0.840)
Δlog GDP exporting region	0.848 (0.404)
Δlog GDP importing region	0.928 (0.290)
R^2	0.5526
No. of Obs.	64

those of Frankel and Romer. They find that a 1 per cent increase in the trade share of GDP induces a 1 per cent increase in income per capita.

2.2.2.1 *Empirical Estimation*

In the subsequent analysis, the hypothesis of increases in the export share of output leading to higher GDP (growth) is tested. For reasons of consistency, the countries chosen for the analysis are the same used in the empirical analysis of the relationship between transport costs and trade volume, namely Germany and Italy.

Most studies focus on developing countries

The present analysis is thus focusing on highly developed countries. This is in contrast to most of the existing literature on trade and economic development which has been concentrating on developing countries. For that reason, differences in empirical results might occur which may not be expected based on existing hypothesis on trade and economic development. These differences might occur because the existing findings on developing countries are not easily applicable to developed countries such as Italy or Germany due to structural differences. In a similar vein, as to the effects of infrastructure investments and their impact on aggregate output differences to previous hypothesis might also occur. There is a wide consensus that in order to *develop* an economy in the first place, infrastructure investment is a necessary precondition for growth, but in countries which are endowed with a well-developed infrastructure the need for further investment to achieve further growth is questionable. The same

Impacts in developed countries may differ significantly

argument might apply to the relationship between trade and output.

Methodology

The methodology that is applied in the present study to examine the relationship between trade and income and income growth respectively is a simple OLS regression. GDP and GDP growth respectively act as the dependent variable while the independent variables are constructed by various measures reflecting the trade relations of the examined country with the rest of the world. All regressions are lagged by one period. The regressions are estimated with robust standard errors.

Data

With a major aim of the present analysis being the examination of the effects of trade on GDP in a **regional setting**, data is collected on a regional basis. The data cover **10 consecutive years from 1994 to 2003**. Export data depict overall exports from every single **Italian and German** region respectively to the world over a period of 10 years between 1994 and 2003.[118] Import data depict overall imports from the world to every single Italian and German region for the same period of time. As previously done in the analysis on the relationship between transport costs and trade, the regions are consolidated into four geographic areas, west, east, south and north. Italian import and export data was obtained from the homepage of the Italian National Statistics Institute ISTAT. GDP data on the Italian regions was obtained from the Chamber of Commerce in Bolzano (Camera di Commercio di Bolzano). German trade and GDP data was obtained from the German National Statistics Institute DESTATIS.

Results

The first regression that was estimated is the following:

118 Italy: Regioni; Germany: Bundesländer

$$\log GDP_i = \alpha + \beta_1 \frac{\exp_w + imp_w}{GDP_i} + \beta_2 \frac{\exp_w - imp_w}{GDP_i} \qquad (2.20)$$

GDP_i … GDP of region i
exp_w … world exports from region i
imp_w … world imports of region i

β_1 denotes the coefficient indicating the strength of the relationship between trade integration and GDP while β_2 denotes the coefficient indicating the strength of the relationship between export surpluses and GDP.

The estimation of this regression produces positive and significant results for the trade integration variable. For the share of net exports variable this regression produces not significant though still positive results. The coefficient of the trade integration variable is much higher with 0.85 than the coefficient of the share of net exports with only 0.07. R^2 is satisfying at 0.4043 explaining 40 per cent of the variance of GDP.

In order to test whether the regression still holds when estimating growth rates, a second regression that was estimated as follows:

$$\frac{GDP_t^i}{GDP_{t-1}^i} = \alpha + \beta_1 \left(\frac{\exp_t^w + imp_t^w}{GDP_t^i} \right) / \left[\frac{\exp_{t-1}^w + imp_{t-1}^w}{GDP_{t-1}^i} \right]$$

$$+ \beta_2 \left(\frac{\exp_t^w - imp_t^w}{GDP_t^i} \right) / \left(\frac{\exp_{t-1}^w - imp_{t-1}^w}{GDP_{t-1}^i} \right) \qquad (2.21)$$

The estimation of regression (2.21) corresponds to the estimation of regression 2.20 in growth rates and produces positive and significant results for the trade integration variable as well as for the share of net exports variable. But the R^2 of 0.1460 casts doubt on the significance of these results. This is similar to the results of Frankel and Romer (1999), who also get positive and significant results for the regression in levels and to Rodriguez and Rodrik (2001), for example, who do not get any positive effects for their estimations in growth rates.

Results give a hint on the relationship between the two variables

It has not been tested for endogeneity bias because it is very hard to find good instruments at the regional level. In this respect it should be conceded that the use of a simple OLS regression without instrumental variables can only give a hint on the relationship between the two variables. Still, the above presented results are very similar in magnitude to those in the literature as well as regarding the fact that the regression in levels is significant in the literature as well as in this study while the regression in growth rates is only significant in a very limited number of studies most of them report insignificant results. This produces similar evidence as the literature indicating that trade integration matters for GDP levels. An increase in the trade share of GDP by 1 per cent leads to an increase in GDP by 0.85 per cent.

2.2.3 Concluding Empirical Remarks

Elasticity of bilateral regional trade flows with respect to transport costs

Results from the estimation of the gravity equation that has been applied to study the effects of changes in transport costs on bilateral regional trade volume between Germany and Italy indicate **a significant negative effect of increases in transport costs on bilateral trade volume**. The elasticity of regional bilateral trade flows between Germany and Italy with respect to actual transport costs between the regions is −2. If transport costs increase by 1 %, bilateral regional trade volume decreases by 2 %. This sensitive reaction of trade flows indicates that transport costs importantly matter for trade flows between developed countries.

Effects of changes in the trade share of output

In a second step, the effect of changes in the trade share of output on output (growth) has been analyzed by means of a regression analysis. The scope of the analysis again is Germany and Italy on a regional level. **Results indicate a significant positive effect for the trade integration variable on output as well as a significant positive effect of the net export share of output on output**. Opposed to this, results from the regression in growth rates do not yield meaningful

Table 2.3 Effects of changes in trade volumes on GDP

Constant	25.15
	(0.19)
log trade share of GDP	0.85
	(0.10)
log net export share of GDP	0.07
	(0.07)
R^2	0.4043
No. of Obs.	198

Table 2.4 Effects of changes in trade volumes on GDP in growth rates

Constant	-0.05 (0.04)
Δlog trade share of GDP	0.61 (0.14)
Δlog net export share of GDP	0.23 (0.06)
R^2	0.1460
No. of Obs.	198

results due to the low R^2. As it is difficult to come up with instrumental variables at the regional level, no instruments have been used. Therefore, the regression might suffer from endogeneity bias, which of course represents a major limitation. At this point the need for further research digging deeper into the relationship between trade shares and output (growth) in developed economies is apparent.

There is a clear link between output and goods transport

Summing up, there is a close link between the conditions to transport goods – represented by transport costs – and bilateral regional trade volumes. Furthermore, trade shares of output seem to matter for output in developed economies such as Germany or Italy. Linking these two results indicates a clear dependence of output on goods transport. GDP still seems to be closely coupled to freight transport. If transport costs increase, trade volumes decrease. If the trade share of output decreases, GDP is likely to follow. Since this empirical analysis certainly has its limitations concerning the number of countries being put under scrutiny and since the available data are not fully satisfying, further research that builds on these results as hypothesis should be made to test the relationship between these variables.

2.3 Concluding Remarks

Previous research focuses on the link between infrastructure investment and output

Previous research on the relationship between transport and economic growth is usually based on the relationship between transport infrastructure investment and output (growth). This link is examined by production or cost function modeling using aggregated data on transport infrastructure investment and aggregated output data. Both approaches usually produce results indicating productivity gains induced by infrastructure investment. Despite variations in magnitude, the general conclusion that can be drawn from previous research on the link between public infrastructure investment and private sector output is that there is a positive impact of public capital on private sector output and thus on economic growth. Apart from specification problems casting doubt on the viability of the results, investment leading to output growth is a tautological relationship. Assuming that production factors are not completely employed investment always leads to increases in output.

In the present study transport has been linked to output through trade

In the present study an innovative way of understanding the mechanisms and effects linking transport to economic output has been chosen. Unlike in previous research endeavors the relationship between transport and economic output has been scrutinized further by linking this relationship to trade. This has been based on the following hypotheses derived from existing literature and theory: Transport enables trade in the first place and can thus be regarded as a precondition for trade to happen. Trade again is supposed to foster economic growth by promoting more efficient production structures and techniques.

These two links have been studied in the present analysis in order to shed further light on the coupling of transport and economic growth. This analysis is structured in the following way: **Firstly, the impact of changes in transport costs on trade volumes is examined in order to be able to estimate the impact of physical trade barriers in developed economies. In a second step, the impact of changes in trade volumes on aggregate output is analyzed**.

Theory proposes a negative effect of trade barriers on trade volume. The magnitude of the physical trade barriers arising from distance between trading partners is reflected in the amount of transport costs which act similarly to tariffs regarding their impact on trade volume. Due to the steady decline in trade barriers arising from tariffs, the relative importance of transport costs on trade volume increases.

In the empirical analysis presented above the impact of changes in transport costs on trade volume between Germany and Italy is examined. **Results indicate a decline of 2.06 per cent in trade volume due to a one per cent increase in transport costs**. The impressive magnitude of these results shows that changes in the transport system may have drastic impacts on trade volume.

How these again affect income and income growth respectively has been examined thereafter. Results presented here support the notion of trade being important for the level of income although the results reflect no clear positive effects of trade on growth. Due to the limited scope and resources of the present research, further efforts should be made relating trade to growth in a regional setting using instrumental variables to control for endogeneity bias which could help us understand the relationship between trade and growth in developed countries more clearly.

The empirical evidence in the present study confirms theoretical predictions

Still, the empirical evidence presented in this study confirms theoretical predictions as well as other empirical evidence in the literature on the relationship between trade and aggregate output. **Results indicate a positive effect of a higher trade share on GDP**. The coefficient of 0.85 of the trade integration variable regressed on GDP resulting from the present analysis is similar to the results presented by Noguer and Siscart (2003) in their very cautious estimation. This indicates that by increasing the trade share of output by 1 % GDP subsequently increases by 0.85 % . Results from Frankel and Romer (1999) who also addressed simultaneity bias are much stronger showing an increase in GDP of 2 % per cent given a 1 percentage increase in the trade share of GDP. Concerning the effects of changes in trade on GDP growth the empirical evidence presented also yields significant effects but the low R^2 casts doubt on the results.

Economic output is closely coupled to freight transport

On the basis of these results for the overarching questions guiding this research endeavor that is, if economic output is (closely) coupled to freight transport, this analysis for Germany and Italy indicates that trade sensitively reacts to changes in transport conditions – represented by transport costs. It also shows that trade shares determine economic output. Therefore it can be concluded that there is a close link between goods transport and economic output. Although this close relationship may hold true for the analyzed period of time and the selected countries, this may not be stable in the future due to possible changes in the economic structure. This represents an important limitation regarding the intertemporal validity of these results.

3 Qualitative Aspects of Transport Infrastructure

Exploring the Importance of Transport Costs Versus Transport Quality for Manufacturing and Distribution Companies

Trade volume acts as an important determinant of GDP

The empirical evidence on the relationship between transport, trade and output presented in the previous chapters indicates that trade volume reacts very sensitively to changes in transport costs and acts as an important determinant of GDP. Thus, changes in transport costs impact output via the trade channel. As freight transport is a necessary precondition for trade to occur in the first place other characteristics of transport services than monetary transport costs such as reliability, speed or flexibility are also likely to have an impact on trade volumes.

Previous studies analyzing the impact of transport on trade volume usually take distance as a proxy for transport costs and thus for the magnitude of the barriers to trade resulting from distance. In this framework changes in distance represent the only influence on transport costs, which leads to the exclusion of qualitative aspects of transport services.

The qualitative aspect of transport services

Bougheas et al. (1997) emphasize that transport costs are to some extent dependent on geographical characteristics but are inversely correlated with the development of infrastructure. The use of distance as the only proxy for transport costs leads to the omission of important variables representing the qualitative aspect of transport services. Various characteristics of developed economies make the plausibility of this restriction even more doubtful as the significance of other transport characteristics than costs for industrial decisions seems to increase steadily.

"We need a representation of the transport sector which recognises the specific transport requirements of individual sectoral users, i.e. transport costs are not just distance related, but relate to bulk, perishability, fragility, value, etc."[119]

Thus, service criteria such as speed, probability of late arrival, extent of late arrival or probability of undamaged arrival are often not taken into account although they represent costs indirectly associated with transporting goods. Overall trans-

119 Vickerman, R.; (1995), p. 227

port costs consist of a direct part e. g. tolls or fuel costs and an indirect part e. g. costs associated with late arrival. For example, time elapsed during shipment increases capital costs or perishability depending on the goods transported. In a same vein, a lack of reliability of arrival may cause production to be interrupted if inventory stocks are not dimensioned for such events. In addition, especially within production and value-creation networks and the development of efficient production techniques such as just-in-time production, nowadays a high reliability of arrival is getting more important. The evolution of production techniques such as just-in-time production makes time especially reliability in terms of punctuality a crucial factor. Congestion tends to become a major, if not the most important, mobility restriction in developed economies disposing of a well-developed transport infrastructure network. The efficiency of the best system starts to rapidly decrease with capacity overloading. Plus, due to the increasing amount of high-end and high-value goods in developed economies, the importance of reliability and time consumed for transport increases. For example Hummels (2001) finds that the willingness-to-pay of exporters for time savings increases, which is more than obvious on the background that he also finds that each day saved due to a shift from sea to air transport is worth 0.5 % of the value of traded goods.[120]

It is crucial to take qualitative factors into account

Due to the obvious importance of these factors (time, reliability, etc.) it is crucial to take them into account in the analysis. I will start by analyzing their relative importance versus each other but also in relation to monetary transport costs by means of a business survey among production and distribution companies. The survey is conducted in the framework of an adaptive conjoint analysis. From the results a factor resembling the relative importance of each item can be derived and thus related to monetary transport costs.

3.1 Previous Research on Transport Quality Attributes

Existing transport research includes studies analyzing qualitative transport choices such as mode choice by means of stated preference techniques.

A study by Danielis et al. (2005) employs a relatively wide focus examining shippers' preferences concerning transport service quality in Friuli-Venezia Giulia, a region in the

120 OECD, (2003b)

North-East of Italy. The ACA methodology was used to conduct face-to-face interviews with logistics managers in order to find out their preferences concerning freight service quality attributes. The sample consists of 65 mostly small or medium sized manufacturing firms in the North-East of Italy, all of which buy transport services from third party providers.

The attributes analyzed consist of costs, travel time, risk of delay and risk of damage and loss. Results show a strong preference for quality attributes such as time, reliability and safety as opposed to costs. They indicate a high willingness to pay for service attributes especially for reliability and safety. Segmentation analysis shows that preferences across regions tend to be rather homogenous while they tend to be heterogenous depending on the nature of the goods shipped and on the firms' characteristics such as size, which is positively correlated with reliability.

It is important to note that the risk of delay is only specified as magnitude of the delay but there is no category denoting the probability of this delay occurring which could be at least as important as the magnitude of the delay. Just consider a shipment that is late for 3 days in 0.001 per cent of the time, while another shipment is late for half an hour in 50 per cent of the time. The second case could have much more severe repercussions on production processes even though the magnitude of the delay is much less. This is one major point that distinguishes the following analysis from the study conducted by Danielis et al. (2005).

Other authors using stated preference techniques in the field of freight transport choices include Maier and Bergman, Bolis and Maggi or Fridstrom and Madslien, all of the edited in the book by Danielis (2002). But most of this research is limited to very specific transport alternatives, like the choice between trucking and rail, intermodal transport along a corridor, the choice between external carrier and own transport, or simply the value of time.

Shinghal and Fowkes (2001) for example examine determinants of mode choice for freight services in India by means of an adaptive stated preference survey. They claim to be the first ones examining freight service attribute valuations in developing countries. The main motivation was to find out if regular domestic container train services between main centers were sustainable. Their sample consists of 32 firms engaged in export and import traffic for various commodities such as auto parts, chemicals, food products etc. They found out that "the frequency of service appears to be an important factor in mode choice, especially for the manufactured goods sectors …"[121] Concerning the scheduled journey times, most sectors require a discount of about 12 % of current costs per day for slower service. Reliability of transit times is reported to be important for exporters as well as for importers of auto parts due to the effect it can have on the production process.

The focus on a very specific objective limits the insight gained to a specific research question in such a way that the general information on people's preferences concerning transport service quality attributes cannot be isolated and transferred to other questions. Regarding the applicability of existing research on travelers' and shippers' preferences on different transport quality aspects, it is not yet extended to an analysis of the effects of different levels of transport quality on actual transport choices, and thus not incorporated into the planning process of infrastructure.

"The development of traffic and transport models is still hampered by a lack of insight into attributes of transport service quality such as reliability, safety and flexibility."[122]

This constitutes additional motivation to conduct research on freight transport service quality attributes in order to be able to shed further light on this still rather obscure field.

3.2 Empirical Analysis

3.2.1 Methodology

Revealed preference methods and stated preference methods

For analyzing the decision making process of managers in firms or individual consumer behavior it can be between revealed preference methods and stated preference methods. Applying revealed preference methods actual behavior on markets is observed and subsequently used in market analysis. Opposed to this, stated preference methods do not rely on the observation of behavior but instead respondents are asked to express their reactions to changes in product characteristics, market conditions etc. This allows the researcher to explore hypothetical or virtual decision contexts and to control relationships between attributes, which permits mapping of utility functions with technologies different from existing ones. Critics state that this advantage in terms of flexibility might be offset by the fact that real behavior is still very different from statements concerning one's actions.

121 Fowkes, T.; Shinghal, N.; (2002), p. 376
122 VanBroeckhoven, B.; Witlox, F.; (2003), p. 3

Danielis and Rotaris (1999) stress the informative advantages of ad hoc stated preference interviews in transport research:

"They provide an essential source of direct, up-to-date information on individual firms' preferences for modal attributes .... Stated preference studies are extremely useful to understand individual firms' micro behavior."

The application of stated preference methods allows the exploration of choice behavior and hypothetical decisions and provides rich insight into attribute trade-off information. Thus, it represents a suitable method to analyze the decision making process among transport managers.

An adaptive conjoint analysis is applied

In the subsequent analysis an adaptive conjoint analysis (ACA), which is a stated preference technique, is used. It is frequently applied in marketing research to study consumer preferences. It evolved from the seminal research of Lucey and Tukey (1964). Their theoretical contributions were applied in practice by a number of psychometricions, including Carroll (1969), Kruskal (1965) and Young (1969). A variety of non-metric models for computing attribute-level values from respondents' preference orderings across multi-attributed stimuli such as descriptions of products or services was developed.[123]

The assumption of utility maximization constitutes the basic framework, whereas the utility of a multi-attribute alternative is decomposed into a set of part-worth utilities. Thus, conjoint analysis is based on the presumption that individuals value a product or service by combining the value provided by the attribute of the product.

"Products or services are thought of as possessing specific levels of defined attributes, and a respondent's 'liking' for a product is modeled as the sum of the respondent's 'utilities' for each of its attribute levels."[124]

Respondents are asked to express their preferences

In the tradition of stated preference methods, "respondents are asked to express their preference, to rate, rank, or choose between hypothetical goods (alternatives) which are described with a set of attributes."[125]

123 Green, P. E. et al.; (2001)
124 Sawtooth Software Inc., (2002)
125 Bouffioux, C.; (2002), p. 3

The strengths of the conjoint analysis

"The strength of the conjoint analysis is its ability to ask realistic questions that mimic the tradeoffs that respondents make in the real world. Respondents evaluate product alternatives (concepts) described by various attributes and indicate which products they prefer. By analyzing the answers, conjoint analysis can estimate the weights and preferences respondents must have placed on the various features in order to result in the observed product preferences."

Opposed to direct questioning methods, the conjoint analysis enables respondents to make difficult trade-offs similar to the ones encountered in the real world. This includes the fact that buyers cannot get all the best features at the lowest price e. g. the fastest, most reliable transport service at the lowest price.

Another advantage "over other stated preference methods is that it confronts the respondent with a simulated decision that is similar to decisions the respondent makes frequently. This reduces the risk of strategic answers by the respondent or that the respondent misunderstands the questions."[126]

Results from conjoint analysis indicate respondents' reactions to changes in attributes or attribute levels. Main applications of the conjoint analysis concern new product/concept evaluation, repositioning, competitive analysis, pricing, and market segmentation.[127]

3.2.1.1 Design of the Questionnaire

Evaluation of the Stimuli

The ACA interview process

The ACA interview consists of different sections each with a specific purpose. It is distinguished between Ratings/Rankings, Importances, Pairs (Trade-off questions) and Calibration Concepts. I will briefly comment on all four concepts and then specify which ones are suitable for the subsequent analysis and why.

1) Preference for Levels: The respondent rates or ranks the levels within attributes for preference. This category is only necessary in cases where it can not be determined a prior whether a lower or higher levels are considered to be better. E. g. consider costs, it is clear a priori that lower costs are better for any respondent, so that this section is

126 Maier et al.; (2002), p. 323
127 Bouffioux, C.; (2002)

redundant. As it is possible to determine the preference for levels a priori for all attributes used in the subsequent analysis this section is excluded in the questionnaire. So the first category included in the questionnaire is the following:

2) Attribute Importance: Having specified preferences for the level within each attribute, the relative importance of each attribute to the respondent is determined. After this section the core of the conjoint analysis follows, the trade-off section.
3) Paired-Comparison (Trade-off Questions): In the first two sections only prior information has been collected, whereas in the pairs section the actual conjoint analysis is performed. The respondent is shown two alternative product concepts, is asked which is preferred and to indicate strength of preference. Firstly, the computer constructs a crude set of estimates for the respondent's utilities from the information gathered in the ranking or rating of levels and the ratings of importance of attributes sections.
4) Calibrating Concepts: Those attributes determined to be most important now constitute a series of "calibrating concepts". The respondent is shown various product concepts, ranging from her most to her least preferred profile, and is asked to estimate the likelihood of buying the product according to the specified concept. This section is also omitted in the questionnaire as it is not important for the subsequent analysis whether the specified transport services will actually be bought or not.

The subsequent ACA consists of attribute importance questions and trade-off questions. The attribute importance section is rather short and merely consists of questions denoting the importance of certain attributes versus others. The setup of the trade-off questions is more complex and what is called an experimental choice design needs to be specified which is described in the following section.

3.2.1.2 Specification of Attributes and Attribute Levels

The number of attributes

At the beginning of performing an adaptive conjoint analysis the attributes describing the good or service in question and its levels need to be specified. This requires a thorough look on the transport service in order to be able to determine the characteristics that are crucial in transport managers' decision processes. On the other hand, one needs to be careful not to pick too many attributes, due to the fact that the inclusion of more than six or seven attributes may threaten the validity of the results. So the number of attributes should be no higher than that but this subset of characteristics should be able to describe all variables considered in transport managers' decision processes.

Attributes analyzed in previous studies

In a report from the Finish Ministry of Transport, the most important requirements of industry concerning transport quality are accounted for by cost efficiency, punctuality and speed. According to the report, punctuality is the most important factor in road transport for trade, industry manufacturing high technology products, food and textile industry. Cost efficiency is stated to be equally important for those sectors but only in sea transport and not in road transport. Speed is especially important in air transport for the trading sector and for industry manufacturing products with high degree of processing. Authors of the report expect speed and punctuality to become more important until the year 2020.[128] The higher value-density of goods transported as the economy moves from low value bulk goods such as coal to high-value industrial products increases the need for fast and reliable transport.[129] A survey by Bolis and Maggi[130] among Italian shippers indicates reliability to be the most important aspect followed by price, speed and safety.

Danielis et al. (2005) use the following four attributes describing transport services: cost, travel time, punctuality and probability of damage or loss. Punctuality is specified as risk of delay of a certain amount of time but the probability of it occurring is not included in the analysis. As depicted above, I think it is crucial to include the probability of the delay. So in the subsequent analysis punctuality will be described by two attributes: the magnitude of the delay in hours and the probability of the delay in percent.

Apart from the literature studied, a set of in-depths interviews[131] with decision makers in the field of transport supplemented the information basis for the selection of the right attributes and their levels. From the information gathered, speed, reliability, punctuality, price, flexibility and safety turned out to be the most relevant attributes describing transport quality.

128 MTC, Finland; (2002)

129 Gaube, V. et al.; (2003)

130 In Danielis, R. ; (2002)

131 Interviews with Mr. Hausmann of BMW Group, Munich and with the managing director of a major international forwarding agent stationed in Austria who wants to stay anonymous.

Attribute levels

Concerning attribute levels Danielis et al. (2005) relate levels of cost and travel time to their current levels so that the respondents need to trade-off improvements in certain attributes against deterioration of other attributes. Cost is specified in percentages deviating from current cost while travel time is denoted in days varying from current travel time. Punctuality is denoted as risk of delay in days, risk of damage and loss in percentages. Specifying risk of delay in days is rather inaccurate. In some industries that are bound to very strict production regimes, that do not allow for delays in between, a delay in the range of a day is unacceptable. Therefore it is necessary to specify the magnitude of the delay in smaller levels such as hours or half hours. Risk of damage and loss is only denoted in three levels, zero risk, risk of 5 % and risk of 10 % . Most industries consider a risk of 5 % as unacceptable, so that the inclusion of the 10 % level is abundant while specification of risk below the 5 % level should be divided into more levels in order to get meaningful results.

Risk of delay is further described by the categories specifying the number of alternative routes as well as the number of alternative modes. The more alternative routes or modes there are, the lower the probability of a delay due to the possibility evading the delay prone routes and modes respectively.

The attributes and their levels specified in the present analysis

According to these considerations I decided to pick the attributes and specify their levels as follows:

Table 3.1 Specified Attributes and Attribute Levels in ACA

Attribute # 1: Average speed	Attribute # 2: Probability of undamaged arrival	Attribute # 3: Cost	Attribute # 4: Flexibility
70 km/h	99.9 %	equal to existing cost	Submission of order at least 2 hours before
60 km/h	97 %	increase of 5 %	at least 6 hours before
50 km/h	below 90 %	increase of 10 %	at least 1 day before
40 km/h		increase of 15 %	at least 2 days before
		increase of 20 %	

Table 3.2 Specified Attributes and Attribute Levels in ACA

Attribute # 5: Number of possible routes	Attribute # 6: Number of possible modes	Attribute # 7: Delay	Attribute # 8: probability of delay
more than 2	more than 2	punctual (+/- 15 min.)	0.1 %
2	2	delay of 30 min.	3 %
1	1	delay of 2 hours	5 %
		delay of 12 hours	more than 10 %
		delay of more than 1 day	

3.2.1.3 *Experimental Choice Design*

Experimental designs, where the attributes and their levels are varied to create choice alternatives, are used to generate the data. A designed experiment allows the manipulation of one or more variables and their levels by the researcher in order to be able to rigorously test certain hypotheses of interest. The manipulated variable is called a "factor" in the experimental design literature, thus, an experimental design can also be called factorial design. More generally, an experimental design refers to the planning process of "which observations to take and how to take them to permit the best possible inferences to be made from the data regarding the hypotheses of research interest."[132]

The conjoint analysis can also be used for the evaluation of hypothetical products and services

As the conjoint analysis can not only be used for the evaluation of real but also for hypothetical products and services, a bundle of product characteristics is not denoted as "product" but as stimulus. Hence, a stimulus describes a certain combination of parameter-characteristics. The specification of the design of the analysis is composed of the definition and determination of the number of the stimuli. Concerning the number of attributes considered to make up a stimu-

132 Louviere, J. J. et al.; (2000), p. 84

lus, it can be distinguished between the full-profile method and the trade-off method. Applying the full-profile method all attributes are considered to make up a stimulus, while the trade-off method only uses two attributes to describe a stimulus.[133]

Determining the number of stimuli

In a second step, the number of stimuli is determined. Each stimulus is characterized by a number of attributes and levels of those attributes. An efficient design should be chosen in the sense that it permits to estimate the parameters of the choice model with maximum precision.

Concerning the number of combinations of attributes and their levels it can be distinguished between a full or complete design and a fractional design. A design in which all possible combinations of attribute levels are shown to each respondent is called a full factorial design. It forms the basis for the derivation of other factorial designs, though it is only used seldomly. When only a subset of treatments is considered, this is called a fractional design.

Full factorial design

For estimating the parameters of general linear models and/or testing hypotheses based on such models, full factorial designs are very attractive due to their statistical properties. They guarantee that all attribute effects of interest are truly independent, which is called "independent by design". This means that it is possible to estimate statistical effects or parameters of interest independently from one another. Interactions occur if preferences for levels of one attribute depend on the levels of a second, for example, if consumers are less sensitive to prices of higher than of lower quality products. In this case, preferences for combinations of price and quality will require this interaction to correctly represent preferences in statistical models. Earlier studies estimating transport service characteristics typically showed interdependences between service quality and price level. This means, that the response to a change in any one attribute, such as price, depends on the values of the other attributes. A strictly additive model tends to over- or under-estimate at the extremes of the utility space. Opposed to this, in the middle of the space additive models tend to predict relatively well.

Generally, there is broad evidence for the fact that interactions exist in many decision rules. Thus, the application of a strictly additive form may be inappropriate in many cases. Despite the fact that the assumptions that must be satisfied for utility functions to be strictly additive are unlikely to be satisfied in many real markets, the more complex an applied

133 Skiera, B. ; Gensler, S.; (2002)

problem, the less practical it is to use designs that provide relatively efficient estimates of all main effects and two-way interactions. The fact that is reassuring in this context is that additive models predict well in attribute regions of greatest interest even if their parameters are biased.[134]

Fractional factorial designs

Using fractional factorial designs, a selection of a particular subset or sample is chosen in order to estimate particular effects of interest as efficiently as possible. Fractional designs only produce unbiased estimates if the assumption of non-significance of interactions is satisfied. In econometrics this problem is called omitted-variables bias. Despite these problems, creating a fractional design can be justified due to the following facts:

- *"main effects typically account for 70 to 90 per cent of explained variance*
- *two-way interactions typically account for 5 to 15 per cent of explained variance*
- *higher-order interactions account for the remaining explained variance".*[135]

Apart from this, Green (1984) presents empirical evidence for the fact that "deterioration of predictive accuracy caused by including additional parameters is greater than the gain in model realism".[136]

Thus, it can be assumed that the application of ACA as a "main effects only" model is appropriate for the empirical research in the present study. A fractional design is applied due to the impractibility of a full factorial design as this would blow up the questioning process in a disproportional way.

The following sample questions from the questionnaire give an insight into the questioning process. They include importance questions as well as trade-off questions from the fractional design.

Importance Questions

Suppose that 2 transport services were identical in all other characteristics, how important would be the following difference:

134 Louviere, J. J. et al.; (2002)
135 Louviere, J. J. et al. ; (2000), p. 94
136 Bouffioux, C. ; (2002), p. 13

Table 3.3 Example of Importance Questions in ACA

Average speed of 70 km/h	Average speed of 40 km/h

x very important
x somewhat important
x not important.

This question is repeated specifying all attributes in their maximum and minimum levels:

Table 3.4 Examples of Importance Questions in ACA

Probability of undamaged arrival of 99.9 %	Probability of undamaged arrival of 90 %
No increase in transport costs	Increase in transport costs of 20 %
Submission of order at least two hours before	Submission of order more than two days before
More than two possible routes	Only 1 possible route
More than two possible modes	Only 1 possible mode
No delay (+/- 15 min.)	Delay of more than 1 day
Probability of delay of 0.1 %	Probability of delay of more than 10 %

Trade-off Questions

Suppose that two transport services were identical in all other characteristics, which one would you prefer?

Table 3.5 Example of Trade-off Question in ACA

Probability of undamaged arrival of 99.9 % Average speed of 40 km/h	Probability of undamaged arrival of 97 % Average speed of 70 km/h

x x x x x x x x x
Strongly prefer left indifferent strongly prefer right

Attributes and attribute levels will be varied according to previous answers given.

Software

The analysis is conducted by using the software package "Adapted Conjoint Analysis v. 5.4" developed by Sawtooth Software Inc. The software generates the trade-off questions individually for each respondent depending on the answers given in the importance questions. This gives way to an important advantage of ACA which consists of the fact that the researcher is not confronted with the difficult trade-off between a long and tiring interview with questionable quality of data, though broad in focus, and a short, concise interview with narrow focus due to the fact that the respondent is only asked in detail about those attributes and levels of greatest relevance to her individually. By adapting the questionnaire for each respondent it is possible to achieve both a broad focus as well as a relatively short interviewing process which usually yields more reliable data than very long interviews.

"ACA's strengths is its ability to investigate many features without overwhelming the respondent with too much information on the computer screen. The term 'adaptive' refers to the fact that the computer-administered interview is customized for each respondent; at each step, previous answers are used to decide which question to ask next, to obtain the most information about the respondent's preferences. ... Questioning is done in an 'intelligent' way; the respondent's utilities are continually re-estimated as the interview progresses, and each question is chosen to provide the most additional information, given what is already known about the respondent's values. An ACA survey includes a series of questions used to first estimate approximate preferences for features, and then later refines them through focused trade-off questions."[137]

Thus, by customizing the interview for each respondent a partial profiles rather than full profiles technique is used, which makes the interviewing process more manageable, concise and efficient. The respondent is not overwhelmed with the full profile of all attributes only a subset (usually 2 to 5) is shown in each question.

3.2.1.4 Sample

Manufacturing and distribution companies in automotives and food

75 manufacturing as well as distribution companies in the food as well as in the automotive sector in Austria were con-

137 Sawtooth Software Inc.; (2002)

tacted, whereas they were picked randomly. Companies were contacted by phone in order to get contacts of the person(s) in charge of logistics at each firm. Interviews were conducted web-based, which means that respondents were supposed to fill out questionnaires on-line from their own computers. A letter was forwarded to the transport managers directly containing the web address of the questionnaire, username and password information. In order to make it more convenient for respondents an e-mail was forwarded to them as well, containing the link so that they only had to click on the link and type in username and password information. By assigning a username and password to every interviewee, the researcher can be sure that only people addressed fill out the questionnaire and do this only once. After several more e-mails and phone calls, asking for the filled out questionnaire once more, a total number of 29 qualified interviews could be retrieved and were subsequently included in the estimation process.

Interviews were conducted web-based

16 of the qualified questionnaires were obtained from the food sector, 13 from the automotive sector. 31 per cent dispose of less than 100 employees, the majority of 55 per cent have between 100 and 500 employees while 14 per cent employ more then 500 people. 21 % of respondents' firms achieve a yearly turnover of more than 100 Mio. Euros, another 21 % state a figure between 50 and 100 Mio. Euros, 41 % between 1 and 50 Mio. Euros and 10 % have a yearly turnover of less than 1 Mio. Euros.[138] Ten companies use just-in-time production, 3 even use the highly sophisticated just-in-sequence technique. The majority of 16 respondents work in outgoing logistics, 8 in incoming logistics and 5 in both. All companies use road transport, a little less than one third uses either ship, plane, rail or multimodal transport schemes as well. The majority of 48 per cent transport packaged, unproblematic goods such as noodles, 24 per cent transport perishable goods, another 24 per cent transport high-quality end products and only one company claims to transport bulk goods.

3.2.2 Results

From the ACA we get two different sets of results. Relative importances give information on the relative importance of the attributes versus each other, while utility values show the utility provided by a certain level of a specified attribute. Relative importances are specified in per cent, thus summing

Relative importances and utility values

138 The remaining firms do not state sales data.

up the values of all attributes yields 100. Utility values are scaled to an arbitrary constant, thus their absolute values do not have any meaning, comparisons between different attributes and their levels need to be undertaken in order to get meaningful information.

3.2.2.1 Average Relative Importances

Calculating relative importances: the range between the lowest and the highest partial utility value

Relative importances can be calculated by using individual partial utilities which indicate the significance of an attribute level in comparison to other levels but give no information on the importance of the attribute itself. To calculate relative importances the range between the lowest and the highest partial utility value of each attribute is needed. By normalizing the range of the partial utilities of an attribute with the sum of all ranges of attributes relative importances are determined. Values of relative importances are calculated as follows:

$$I_i^m = \frac{(x_{i\max}^m - x_{i\min}^m)}{\sum_{i=1}^{n}(x_{i\max}^m - x_{i\min}^m)} \tag{3.1}$$

I_i^m ... denotes relative importance of attribute i for respondent m
$x_{i\max}^m$ ... denotes maximum partial utility value of attribute i for respondent m
$x_{i\min}^m$ ... denotes minimum partial utility of attribute i for respondent m
n ... denotes the number of attributes

The sum of all importances adds up to one hundred for each respondent so that the values can be interpreted as percentages. In the following table average importances as mean values over all respondents, over all respondents from consumer goods industry and over all respondents in the automotive sector respectively are reported:

Table 3.6 Average Importances from ACA

Attributes	Average Importances		
	Both Industries	Consumer-goods industry	Automotive industry
Average Transportation Speed	9.53	8.31	11.02
Probability of Undamaged Arrival	18.48	18.28	18.73

Table 3.6 *(continued)*

Attributes	Average Importances		
	Both Industries	Consumer-goods industry	Automotive industry
Transport Costs	18.61	17.36	20.14
Flexibility	8.91	9.02	8.78
Number of Alternative Routes	6.57	7.70	5.17
Number of Alternative Modes	6.91	7.50	6.19
Magnitude of Delay	16.94	17.71	15.99
Probability of Delay	14.05	14.12	13.96

The results show that the transport service quality attributes can be put into two categories: Those of rather high importance, ranging between 19 and 14, and those of lower importance ranging between 7 and 10. Transport Costs, Probability of Undamaged Arrival, Magnitude of Delay as well as Probability of Delay belong to the first group, while Average Transportation Speed, Flexibility, Number of Alternative Routes and Number of Alternative Modes are among the second group.

Transport service quality attributes can be put into two categories

A very interesting finding constitutes the fact that there is only a very small difference in results between the two sectors. Average importances are very similar for all attributes regardless of the sector in which the industry is operating.

There is only a small difference between the two sectors

The average importances represent aggregates over all respondents, over the consumer goods sector and the automotive sector respectively. An analysis of individual importance values is presented in chapter 3.2.3, whereas a cluster analysis is performed in order to be able to discern certain groups of respondents with similar values.

3.2.2.2 Average Partial Utility Values

Relative importances provide information on the importance of a certain attribute but they do not give any insight regarding the utility associated with a certain attribute level. For the optimization of a system on the whole, in this particular case

Elasticities of substitution between attributes to compare changes in levels of different attributes

the transport infrastructure system, exactly this kind of information is needed in order to be able to estimate certain trade-offs according to their impact on utility values of users. Elasticities of substitution between attributes need to be calculated in order to be able to compare changes in levels of different attributes.

For example consider the following trade-off: if probability of delay is lowered by 2 per cent while transport costs rise by 5 % does this pose a worsening or an improvement of utility? To be able to answer questions like this partial utility values of the levels are needed in order to be able to compare a worsening in the level of one attribute versus an improvement in the level of another attribute.

Utility values are obtained on an individual basis. This means that we get 29 utility values for each level of the attributes. The total number of levels of attributes is 31, so we get 29 rows, one for each respondent, and 31 columns, one for each variable describing the utility value of each specified level of each attribute. Individual values are of interest for an analysis on a disaggregate level, for example a cluster analysis, which divides respondents into categories according to their utility profile.

Comparing impacts of changes in levels of different attributes

In order to compare impacts of changes in levels of different attributes on utility average values are needed. Before being able to make comparisons partial utility values need to be normalized, then the mean over all respondents can be taken. As already mentioned above, absolute values do not have any expressiveness due to the fact that

"Conjoint utilities are scaled to an arbitrary additive constant within each attribute and are interval data. The arbitrary origin on the scaling within each attribute results from dummy coding in the design matrix."[139]

Normalized partial utility values are obtained as follows:

$$U^m_{x_{ir}} = \frac{(x^m_{ir} - x^m_{i\min})}{(x^m_{i\max} - x^m_{i\min})} * I^m_i \tag{3.2}$$

U^m_{ir} ... denotes normalized partial utility of level r of attribute i for respondent m

x^m_{ir} ... denotes partial utility of level r of attribute i for respondent m

139 Sawtooth Software; (2002), p. 393

I_i^m ...denotes relative importance of attribute i for respondent m
$x_{i\max}^m$ ... denotes maximum partial utility value of attribute i for respondent m
$x_{i\min}^m$ ... denotes minimum partial utility of attribute i for respondent m

Normalized average partial utility values are reported in the following table:

Table 3.7 Average Normalized Partial Utility Values from ACA

Attribute	Levels	Average Utilities	Consumer Goods S.	Automotive S.
Average Transportation Speed	70 km/h	8,82	8,06	9,68
	60 km/h	6,54	5,81	7,27
	50 km/h	3,15	2,52	3,80
	40 km/h	0,15	0,00	0,29
Probability of Undamaged Arrival	99.9 %	18,22	18,57	18,31
	97 %	10,86	10,75	10,82
	below 90 %	0,02	0,03	0,00
Transport Costs	no change	17,79	17,59	18,68
	increase of 5 %	13,39	12,89	14,28
	increase of 10 %	9,72	7,59	12,88
	increase of 15 %	4,48	4,25	5,38
	increase of 20 %	0,75	1,22	0,14
Flexibility	submission of order	7,63	7,56	7,43
	at least 2 hours before	5,72	5,31	5,89
	at least 6 hours before	3,82	3,54	3,86
	at least 1 day before			
	more than 2 days before	0,64	1,17	0,00
Number of Alternative Routes	more than 2 alternate routes	6,53	7,61	5,50
	2 alternative routes	3,31	3,88	2,75
	only 1 possible route	0,06	0,11	0,00

Table 3.7 *(continued)*

Attribute	Levels	Average Utilities	Consumer Goods S.	Automotive S.
Number of Alternative Modes	more than 2 alternate modes	6,42	7,61	5,89
	2 alternate modes	4,14	4,64	3,85
	only 1 possible mode	0,11	0,00	0,21
Magnitude of Delay	punctual	15,15	15,19	13,82
	(+/15 min.)	14,12	14,32	12,95
	delay of 30 min.	10,97	10,84	10,38
	delay of 2 hours	3,92	4,50	3,96
	delay of 12 hours	0,93	1,03	0,70
	delay of more than 1 day			
Probability of Delay	0.1 %	13,59	13,21	13,85
	3 %	8,64	8,09	8,94
	5 %	5,69	5,78	4,85
	More than 10 %	0,54	0,62	0,98

Comparing normalized partial utility values

Now normalized partial utility values can be compared easily. The improvement from one level to another "brings" additional utility while a worsening from one level to another of a different attribute "costs" utility. By subtracting the costs from the gains we get net partial utilities of changes in levels of different attributes.

$$\Delta U = \Delta U_i + \Delta U_j \tag{3.3}$$

For example, a decrease in the magnitude of delay from two hours to zero corresponds to a gain in partial utility by 4.18 while an increase in transport costs by 5 per cent corresponds to a loss of partial utility by 4.4, so there is a minor loss of net partial utility of 0.22.

Utility changes corresponding to a one percent, one hour or one km/h change in attribute level can be calculated. Subsequently, elasticities of substitution between attributes can be derived, which allows us to determine the necessary improvement in one attribute to off-set a deterioration in another attribute in order to be able to maintain the same utility level. Generally, the elasticity of substitution between changes in attribute *i* and attribute *j* can be calculated as follows:

$$e_{ij} = -\frac{\Delta_{\%}U_{jp}}{\Delta_{\%}U_{ip}} \tag{3.4}$$

e_{ij} ... denotes elasticity of substitution between attribute i and attribute j
$\Delta_{\%}U_{jp}$ ... denotes change in utility due to a 1 per cent change in attribute j
$\Delta_{\%}U_{ip}$... denotes change in utility due to a 1 per cent change in attribute i

Elasticity of substitution between transport costs and probability of delay

For example, how does a 1 % increase in transport costs, starting from today's level, compare to a 1 % decrease in probability of delay? The one per cent increase in transport costs corresponds to a utility loss of 0.88, while the one per cent decrease in probability of delay leads to an increase in utility of 1.7. This yields an elasticity of substitution between transport costs and probability of delay of –2.13.

3.2.3 Cluster analysis

Subsuming respondents into homogenous subgroups

In order to be able to subsume respondents with similar utility profiles into homogenous subgroups a cluster analysis, also called segmentation analysis, is performed. It seeks to identify a set of groups which both minimize within-group variation as well as maximize between-groups variation.

Hierarchical clustering

Using hierarchical clustering every case is initially considered a cluster, then two cases with the lowest distance or highest similarity are merged into a cluster according to a specified selection criterion. Hierarchical clustering allows the researcher to select a measure of distance which is most appropriate. The most common distance measure is the Euclidean distance. Using Euclidean distance the smaller the distance the more similar the cases, but it is only appropriate for comparison across variables that do not have very different variances. Due to the fact that the results from the ACA have been standardized, this problem is evaded and the Euclidean distance is used as distance measure in the following hierarchical cluster analysis.[140]

As selection criterion the Ward method is used, which merges cases into clusters by minimizing the variance within a cluster. Two cases are joined into a cluster if this merger results in the minimum increase in the error sum of squares.

140 Backhaus, K. et al.; (2000)

Thus, at each stage the average similarity of the cluster is measured. A case is selected to join the cluster if its inclusion in the cluster produces the least increase in the sum of squared deviations.[141]

There are three clusters

The number of clusters was specified at three due to the relatively high similarity in respondents' answers. By increasing the number of clusters, only very small changes in average relative importance at each cluster could be found. Cluster 1 contains the major part of respondents with 14, cluster 2 is second with 10 respondents and cluster 3 includes 5 respondents. No outliers were found so all respondents are included in one of the three clusters.

141 Field, A.; (2000)

Table 3.8 Relative Importances of Cluster 1

Relative Importance	Attribute	Per cent
1	Probability of Undamaged Arrival	20.13 %
2	Magnitude of Delay	18.37 %
3	Transport Costs	15.26 %
4	Probability of Delay	12.49 %
5	Flexibility	11.44 %
6	Average Speed	8.49 %
7	Number of Alternative Routes	7.13 %
8	Number of Alternative Modes	6.68 %

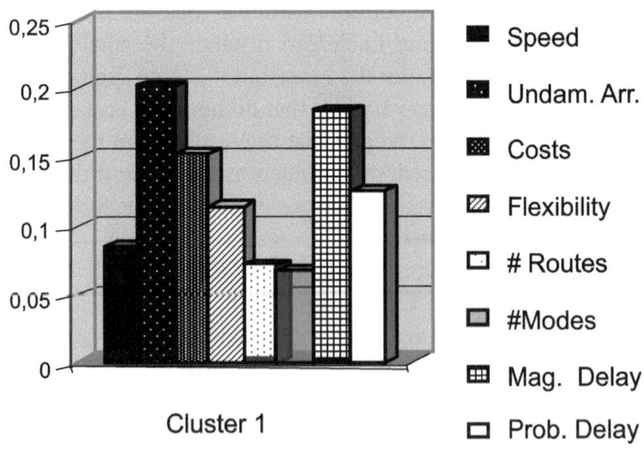

Fig. 3.1 Cluster 1

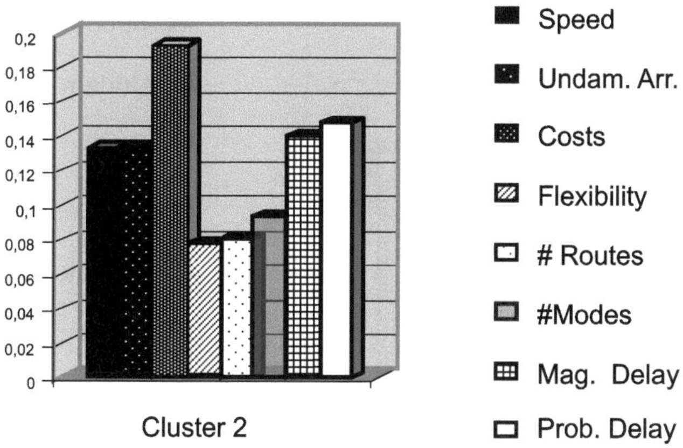

Fig. 3.2 Cluster 2

Table 3.9 Relative Importances of Cluster 2

Relative Importance	Attribute	Per cent
1	Transport Costs	19.26 %
2	Probability of Delay	14.77 %
3	Magnitude of Delay	14.02 %
4	Probability of Undamaged Arrival	13.38 %
5	Average Speed	13.31 %
6	Number of Alternative Modes	9.39 %
7	Number of Alternative Routes	8.03 %
8	Flexibility	7.83 %

On the y-axis relative importances are drawn. Cluster 1 shows a very high sensitivity to the probability of undamaged arrival variable with 20.13 % of overall importance of all variables. Magnitude of delay ranges second at 18.37 % and transport costs come third with 15.29 % . Opposed to the relatively high importance of the magnitude of delay variable, the variables specifying the number of alternative routes or modes rank last with around 7 % each.

Respondents in cluster 2 are the ones reacting most sensitively to transport costs ascribing them a relative importance of 19.26. The delay variables rank second in importance with probability of delay at 14.77 per cent and magnitude of delay at 14.02 per cent, closely followed by probability of undamaged arrival at 13.38 % .

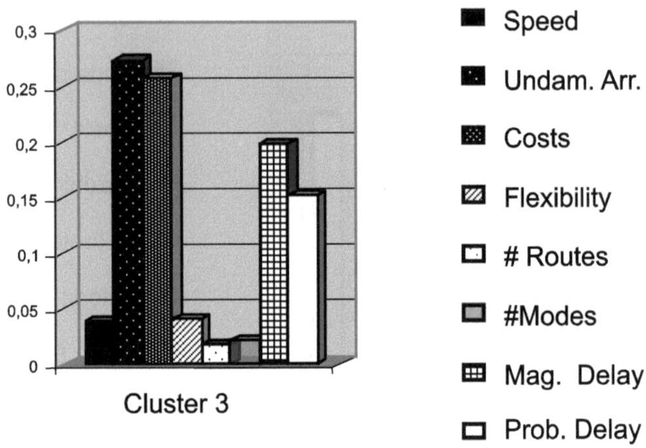

Fig. 3.3 Cluster 3

Table 3.10 Relative Importances of Cluster 3

Relative Importance	Attribute	Per cent
1	Probability of Undamaged Arrival	27.34 %
2	Transport Costs	25.79 %
3	Magnitude of Delay	19.76 %
4	Probability of Delay	15.19 %
5	Flexibility	4.01 %
6	Average Speed	3.96 %
7	Number of Alternative Modes	2.10 %
8	Number of Alternative Routes	1.85 %

Relative importances vary most among respondents in cluster 3. The most important two variables, probability of undamaged arrival and transport costs, account for more than 53 per cent of overall importance, with 27.34 and 25.79 per cent respectively.

Respondents from both branches almost equally distributed among the three clusters

After the formation of clusters, I look at the structure of each one of them regarding the respondents' characteristics such as branch or production technique. Concerning branches no particular pattern was discernible so that respondents from both branches are almost equally distributed among the three clusters. Average distance traveled by the goods transported does not account for a differentiation criterion for the cluster classification. Regarding production technique, 70 % of all respondents using just-in-time techniques are in cluster 1, the other 30 % are in cluster 2.

Formation of clusters regarding normalized partial utilities

Concerning the composition of the clusters, the formation of clusters regarding normalized partial utilities yields very similar results to the clusters formed according to relative importances which are reported above. Cluster 3 contains the same respondents as cluster 3 formed according to relative importances. Cluster 1 contains the same respondents except for 3 respondents which corresponds to 10 % of all respondents. Cluster 2 differs in two respondents or 7 % .

3.3 Combination of Results with Traffic Flow Data

Using elasticities of substitution for optimizing the use of transport infrastructure systems

The results from the ACA provide estimates of elasticities of substitution between direct monetary transport costs and transport service quality attributes. In order to be able to use these estimates for the optimization of transport infrastructure systems additional information concerning the relation between density and transport service quality attributes is needed. Concerning the transport service quality attributes, probability and magnitude of delay, which have been identified in the previous chapter to be among the most important factors determining transport quality for manufacturing and distribution companies, the stability of flow rates represents a crucial factor affecting their levels.

Traffic flow analysis provides speed-flow relationships describing average speed as a function of density. Macroscopic variables, usually average speed as km/h over all vehicles or vehicles per hour and vehicles per km per lane, are used, with speed or vehicles per hour being the dependent variable. The percentage of freighters is also accounted for due to its different impact on average speed.[142]

Following the assumption that there is no speed limit, average speed for passenger cars on a two-lane, one-way road can by calculated from flow rates with the following formulas[143]:

$$V_{PKW} = 138{,}6\, km/h - 8 * \exp(0{,}235 * s) - 0{,}1 * \exp(1{,}643 * 10^{-3} * (Q_{PKW} + 2Q_{LKW})) \quad (3.5)$$

142 Schick, P.; (2003)
143 FGSV, 1997

If flow rates supersede 3900 vehicles per hour, the following formula should be applied:

$$V_{PKW} = \coth((Q_{PKW} + Q_{LKW}) - 3880{,}52 * 10^{-3}) + 18{,}88\,km/h \tag{3.6}$$

For freighters at flow rates < 650 vehicles per hour the following formula is applied:

$$V_{LKW} = 86{,}1\,km/h - 6 * \exp(0{,}248 * s) - 0{,}1 * \exp(9{,}218 * Q_{LKW}/1000) \tag{3.7}$$

and at > 650 vehicles per hour

$$V_{LKW} = \coth((Q_{LKW} - 645{,}26)/100) + 18{,}93\,km/h \tag{3.8}$$

V_{LKW} … average speed freighters
V_{PKW} … average speed passenger cars
Q_{LKW} … flow rate freighters
Q_{PKW} … flow rate passenger cars
s … slope in %

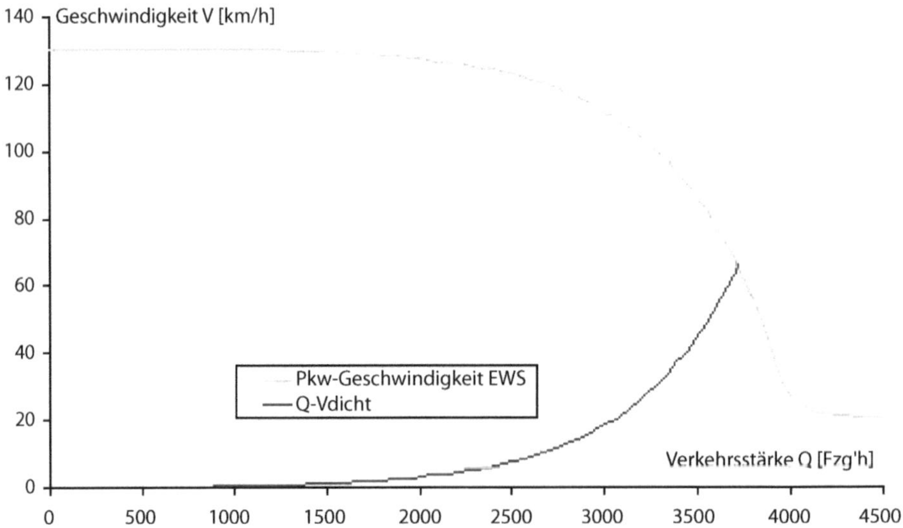

Fig. 3.4 Relationship between flow rate (x-achsis) and speed (y-achsis)
Source: FGSV, 1997

Ideally, which means that everybody is driving more or less at constant speed and minimum required safety distance, traffic follows the grey line. But in practice traffic breaks down much earlier following the inverse black line.

Taking measures to improve efficiency

Concerning the efficiency of measures improving traffic flows, this graph shows that it is important to take into account the relationship between speed and flow rate. Action should be taken at points where the slope of the curve is relatively steep otherwise improvements in transport service quality will be low compared to the efficiency loss due to regulatory action. In order to determine the degree of action that should be taken elasticities of substitution between transport service quality attributes need to be considered. If preferences for increased speed are high measures should be taken relatively early. If preferences for increased speed are somewhat lower but still at a considerable rate action to increase flow rates should be taken at higher flow rates but still only at a point where the slope is relatively steep.

As an example I have looked at actual traffic flow data over the Brenner-Pass. It shows that even during the morning hours traffic does not reach the critical point of 2000

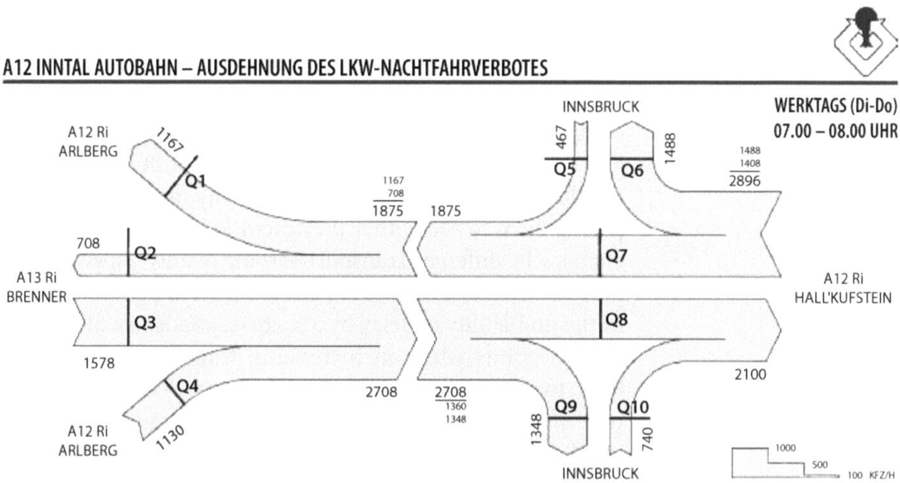

QUERSCHNITT	Q1	Q2	Q3	Q4	Q5	Q6	Q7	Q8	Q9	Q10
KFZ	1.167	708	1.578	1.130	467	1.488	1.408	1.360	1.348	740
LKW (>7,5t) = L75	120	174	172	155	35	45	259	296	31	34
ANTEIL L75	10%	25%	11%	14%	7%	3%	18%	22%	2%	5%
KFZ ohne LKW (>7,5t) = RVK	1.047	534	1.406	975	432	1.443	1.149	1.064	1.317	706
ANTEIL RVK	90%	75%	89%	86%	93%	97%	82%	78%	98%	95%

Flows at morning peak hours (Tuesday through Thursday)

Fig. 3.5 Traffic flow rates at the A12-Brenner-Autobahn interjunction in Innsbruck
Source: Amt der Tiroler Landesregierung, 2005

vehicles per hour. But the graph below shows that the point where the Inntal-Autobahn connects with the Brenner-Autobahn is a bottleneck, during the morning hours traffic reaches flow rates well above 2000 vehicles per hour. Still, the data depict traffic flow rates during the busiest hours of the week, between 7 and 8 am in the morning between Tuesday and Thursday.

3.4 Conclusions

Preferences for transport service quality are substantial

To conclude on the interviews in the context of the conjoint analysis among transport managers exploring the importance of transport quality aspects there is an indication that preferences for transport service quality are substantial. As production techniques become more refined, as the value-added process is increasingly split-up and as thus companies are more and more dependent on high quality transport services the quality aspect of these services gains in importance.

Results of the present study indicate that in terms of transport service quality the importance of reliability is comparable to the importance of direct monetary transport costs (see Table 3.6). Transport managers are mainly interested in goods transport being reliable that is of course goods arriving undamaged but also in time at their destination. By comparing utility values of changes in the different quality aspects, one can estimate their impact on utility. Though the absolute values of utility do not have any meaning, by comparing them to each other they provide information on how changes in different transport service quality aspects compare to changes in transport costs. For example, an increase in the probability of delay by 3 % corresponds to a utility loss of 4.95 points while an increase in transport costs by 5 % leads to a decrease in utility by 4.4 points. While the magnitude of delay plays an equally important role as the probability of delay, flexibility in terms of the need of planning transport services in advance or the possibility to use a different route are less important. Average transportation speed which can be regarded as a proxy for time elapsed during transport is reported to be much less important than magnitude or probability of delay. Thus, the problem with delays obviously seems to be their unexpected nature. If they could be predicted and taken into account when planning production or stocks their repercussions would be substantially lower and easier to deal with for companies.

Measures concerning transport infrastructure investment und use

To provide an outlook on the basis of the results of this empirical analysis, it seems to be worthwhile to take into account user preferences representing the demand side, as well as the actual traffic situation reflecting the supply side with respect to quality of transport services when improving transport infrastructure. Especially delivering predictions of traffic situations on the basis of historical data and up-to-date information on traffic situations combined with surveys incorporating transport managers might be interesting. By combining information on user preferences with traffic flow data which provides information on the relation between infrastructure capacity, infrastructure use and transport service quality such as the probability of congestion, investments in as well as regulation of transport networks can be designed according to the actual needs of users and the particular characteristics of the transport link concerned. Investments should be directed to links that frequently reach the point where the relation between speed and flow rate decreases rapidly due to overuse. If the use of transport infrastructure is regulated, regulation should be applied if speed as well as flow rates can be positively influenced by political actions. Apart from this, the application of intelligent transport information systems like satellite-based navigation can help improve the situation by providing critical information to users in order to increase predictability of the traffic situation. Also intermodal solutions could be applied to help unbundling traffic congestion and to counter delays in goods transports. This combination of information on user preferences with traffic flow data opens up interesting ways for improving the efficiency of transport infrastructure use. In this context, flow rates are not only crucial in determining magnitude and probability of delay but are the primary determinant of road capacity.

The empirical results suggest future political regulation of transport infrastructure use should be designed taking into account user preferences. The existing analysis has shown that predictability, punctuality and other aspects of transport infrastructure quality are very important for enterprises, thus future investment policy and regulation should be designed to contribute to these very aspects which are seen by the users of transport infrastructures, especially the commercial users, as being highly relevant for their business activities. If user preferences are not taken into account this may induce negative effects on economic growth potentials. Therefore designing growth-oriented transport policies and investment decisions concerning transport infrastructures and regulations limiting their use should importantly consider user preferences to be more beneficial in the long term.

4 Concluding Remarks and Outlook

Confronted with the pressing question for policy-makers, decision-makers and scholars if transport growth is really coupled to economic growth, the present study set out to tackle two different sets of questions:

1. What is the relationship between transport, trade and economic growth? How is trade affected by changes in transport conditions and subsequently what are the repercussions on economic growth?
2. Given the empirical evidence, that indicates a close link between transport, trade and economic output, it is in a next step important to learn how to direct transport infrastructure networks towards their most efficient use. Therefore, the question is which characteristics of transport services are essential for companies to organize their logistics operations in an efficient way?

Transport has been linked to economic growth through trade

To come up with answers to these different sets of questions in the present study different forms of analysis have been combined. In order to shed light on the relation between transport, trade and economic growth in an econometric analysis transport has been linked to economic growth through trade. Following the hypothesis that trade is supposed to be influenced by the conditions to transport goods and to act as an important determinant of output different relationships in the triangle of transport, trade and economic growth have been scrutinized.

Trade volumes very sensitively react to changes in transport costs

In the first part of the econometric analysis effects of changes in transport costs on bilateral regional trade volumes are examined by means of a gravity equation. Actual transport costs reported by freight forwarding agencies are used as a proxy for the physical barrier to trade represented by distance. Results indicate that trade volumes very sensitively react to changes in transport costs with an elasticity of −2. If transport costs increase by 1 % trade volumes decrease by 2 %.

Effects of changes in the trade share of output are significant and positive

In a second step, the effect of changes in the trade share of output on output (growth) has been analyzed by means of a regression analysis. Results indicate a significant positive effect for the trade integration variable on output as well as a significant positive effect of the net export share of output on output. Opposed to this, results from the regression in growth rates do not yield meaningful results.

To come up with insights for the second set of questions and to contribute to pressing issues in transport policy, the second part of the study sets out to examine user preferences concerning the importance of quality aspects of transport services. This analysis with a microeconomic perspective has been done to find out more about the mainly overlooked issue of user preferences towards transport infrastructures which should play a more important role in policy design and academic research. In this context, the present analysis sets out to deliver more insights on the relative importance of qualitative characteristics such as reliability or flexibility with respect to monetary transport costs. These aspects have been analyzed by means of an adaptive conjoint analysis surveying logistic departments of manufacturing and distribution companies.

The importance of transport service quality is comparable to the importance of direct monetary costs

Results indicate that in terms of transport service quality the importance of reliability is comparable to the importance of direct monetary transport costs (see Table 3.6). Transport managers are mainly interested in goods transport being reliable that is of course goods arriving undamaged but also in time at their destination. By combining information on user preferences with traffic flow data which provides information on the relation between infrastructure capacity, infrastructure use and transport service quality such as the probability of congestion, investments in as well as regulation of transport networks can be designed according to the actual needs of users and the particular characteristics of the transport link concerned.

Summing up, there is a close link between the conditions to transport goods – represented by transport costs – and bilateral regional trade volumes. Furthermore, trade shares of output seem to matter for output in developed economies such as Germany or Italy. Linking these two results indicates a clear dependence of output on goods transport. GDP still seems to be closely coupled to freight transport. If transport costs increase, trade volumes decrease. If the trade share of output decreases, GDP is likely to follow. Still, a major limitation to the results is that they are viable only short term. If structural changes take place within the economy they may loose part of their significance. Therefore, an analysis on possible structural changes in the future and their impacts on the relation between transport and economic output could provide further insight into the future prospects of goods transport and its meaning for economic output.

For the time being, in order to optimize the use of transport infrastructure with respect to economic efficiency user preferences should be taken into account when deciding on investments in infrastructure or on regulation concerning the use of transport infrastructure networks. The existing analysis has shown that predictability, punctuality and other aspects of transport infrastructure quality are very important for enterprises, thus future investment policy and regulation should be designed to contribute to these very aspects.

Taking stock of the challenges this study could only contribute modestly with raising some issues and delivering some data - limited in its validity, it may be. Hopefully, in future research endeavors some of these aspects are taken up thus contributing with further insight to growth-oriented transport policies.

References

Aberle, G.; (2003), "Transportwirtschaft", 4. Aufl., Oldenbourg Verlag, München.

Aghion, P.; Howitt, P.; (1999), "Endogenous Growth Theory", MIT Press, Cambridge, MA.

Alberto, P.; (2000), „The logistics of industrial location decisions: an application of the Analytic Hierarchy Process Methodology", International Journal of Logistics: Research and Applications, Vol. 3, No. 3, 273 - 289.

Alcalá, F.; Ciccone, A.; (2003), "Trade, Extent of the Market, and Economic Growth 1960–1996", Working Paper, Universitat Pompeu Fabra, Barcelona.

Amt der Tiroler Landesregierung, Abteilung Verkehrsplanung, (2005), "Verkehr in Tirol - Bericht 2004", http://www.tirol.gv.at/themen/verkehr/service/verkehrsbericht/

Anderson, J.E.; (1979); "A Theoretical Foundation for the Gravity Equation", American Economic Review, 69 (1), p. 106 - 116.

Anderson, J.E.; van Wincoop, E.; (2001), "Gravity with Gravitas: a solution to the border puzzle", Working Paper No. 8079, National Bureau of Economic Research, Cambridge, MA, USA.

Arrow, K.J.; Kurz, M.; (1970), "Public Investment, the rate of return, and optimal fiscal policy", Johns Hopkins Press, Baltimore.

Aschauer, D.A.; (1989a), „Is public expenditure productive"?, Journal of Monetary Economics 23, 177 - 200 in Crihfield, J. B., Panggabean M. P. H., (1995), "Is public infrastructure productive? A metropolitan perspective using new capital stock estimates", Regional Science and Urban Economics 25, 607 - 630.

Aschauer, D.A.; (1989b), "Public investment and productivity growth in the Group of Seven", Economic Perspectives 13, 17 - 25.

Aschauer, D.A.; (1991), „Infrastructure: America's Third Deficit", Challenge March/April, 39 - 45.

Aschauer, D.A.; (1997a), "Do states optimize? Public capital and economic growth", The Jeromy Levy Economics Institute of Bard College, Working Paper No. 189.v

Aschauer, D.A.; (1997b), "Output and employment effects of public capital", The Jerome Levy Economics Institute of Bard College, Working Paper No. 190.

Aschauer, D.A.; (1998), "How big should the public capital stock be? The relationship between public capital and economic growth", The Jerome Levy Economics Institute of Bard College, Public Policy Brief Series No. 43.

Ayres, R.U.; Weaver, P. M.; (1998), "Eco-Restructuring: Implications for Sustainable Development", United Nations University Press, Tokyo, Japan.

Backhaus, K. et al.; (2000), "Multivariate Analysemethoden: Eine anwendungsorientierte Einführung", 9. Aufl., Springer, Berlin.

Badri, A.M. et al.; (1995), "Decision support models for the location of firms in industrial sites", International Journal of Operations & Production Management, Vol. 15 No. 1, 50 - 62.

Baier, S.L.; Bergstrand, J. H.; (2001), „The growth of world trade: tariffs, transport costs, and income similarity", Journal of International Economics 53, p. 1 - 27.

Banister, D.; Berechman, Y.; (2000), „Transport investment and economic development", UCL Press, London.

Banister, D.; Berechman, Y.; (2001), "Transport investment and the promotion of economic growth", Journal of Transport Geography 9, 209 - 218.

Bardhan, K. P.; (1970), "Economic Growth, Development, and Foreign Trade", Wiley-Interscience, New York.

Barro, R.J.; Sala-I-Martin, X.; (1995), "Economic Growth", Mc Graw-Hill, New York.

Batten, D.F.; Karlsson, C.; (1996), „Infrastructure and the Complexity of Economic Development", Springer Verlag, Berlin - Heidelberg.

Bernard, A.B. et al.; (2003), "Falling Trade Costs, Heterogeneous Firms and Industry Dynamics", Centre for Economic Performance, London School of Economics, London, UK.

Bernard, A.B. et al.; (2000), "Plants and Productivity in International Trade", NBER Working Papers No. 7688.

Berndt, E.R.; Hansson, B.; (1992), „Measuring the contribution of public infrastructure capital in Sweden", Scandinavian Journal of Economics 94, 151 - 168.

Boarnet, M.G.; (1996), "The direct and indirect economic effects of transportation infrastructure", Working Paper UCTC No. 340, Department of Urban and Regional Planning and Institute for Transportation Studies, University of California, Irvine.

Boisso, D. et al.; (2000), "Productivity and efficiency in the US: effects of business cycles and public capital", Regional Science and Urban Economics 30, 663 - 681.

Bond, E.W.; (1997), "Transportation Infrastructure Investment and Regional Trade Liberalization", Department of Economics, Penn State, University Park, PA.

Bouffioux, C.; (2002), "Conjoint Analysis of Stated Preference Marketing Literature", SSTC project CP-TR-03, Services Federaux des Affaire Scientifique, Techniques et Culturelles, Brussels.

Bougheas, S. et al.; (1997), "Infrastructure, Specialisation and Economic Growth", Keele University Working Paper.

Bougheas, S. et al.; (1999), "Infrastructure, transport costs and trade", Journal of International Economics, 169 - 189.

Bridgman, B.; (2003), "Energy Prices and the Expansion of World Trades", Louisiana State University.

Bruinsma, F.; (1995), "The impacts of new infrastructure on the spatial patterns of economic activities", Serie Research Memoranda No. 12/1995, University of Amsterdam.

Brülhart, M.; (1998), "Economic Geography, Industry Location and Trade: The Evidence", The World Economy, Vol. 21.

Canning, D.; (1989), "The contribution of infrastructure to aggregate output", Department of Economics, The Queen's University of Belfast, UK.

Carroll, J.D.; (1969), "Categorical Conjoint Measurement", Meeting of Mathematical Psychology, Ann Arbor, MI, USA.

Chen, N.; (2004), "Intra-national versus international trade in the European Union: why do national borders matter?", Journal of International Economics, 63, p. 93 - 118.

Cohen, J.P.; Morrison, C. J.; (2001), „Public Infrastructure Investment, Costs, and Inter-State Spatia Spillovers in U.S. Manufacturing: 1982 - 96", Department of Agricultural and Resource Economics, University of California, Davis.

Combes, P.-P.; Lafourcade, M.; (2003), "Core-Periphery Patterns of Generalized Transport Costs: France, 1978 - 1998", Discussion Paper No. 3958, Centre for Economic Policy Research, London, UK.

Conlon, R.M;. (1982), "Transport Cost and Tariff Protection of Australian Manufacturing", Economic Record, Vol. 58, No. 160, p. 73 - 81.

Conrad, K.; Seitz, H.; (1994), "The public capital hypothesis: The case of Germany", Recherches Economique de Louvain, Vol. 54, p. 287 - 307.

Conrad, K.; Seitz, H.; (1997), "Infrastructure Provision and International Market Share Rivalry", Regional Science and Urban Economics, No. 27, p. 715 - 734.

Crihfield, J.B.; Panggabean, M.P.H.; (1995), "Is public infrastructure productive? A metropolitan perspective using new capital stock estimates", Regional Science and Urban Economics 25, 607 – 630.

Crook, G.; (2002), "Infrastructures Framework for Trade", UNCTAD/SITE.

Cunat, A.; Maffezzoli, M.; (2003), "Can Comparative Advantage explain the Growth of US trade?", Working Paper No. 241, IGIER, Bocconi University, Milan, Italy.

Danielis, R.; Rotaris, L.; (1999), "Analysing Freight Transport Demand Using Stated Preference Data: A Survey and a Research Project for the Friuli-Venezia Giulia Region", Trasporti Europei, n. 13.

Danielis, R. (ed.); (2002); "Domanda di trasporto merci e preferenze dichiarate – Freight Transport Demand and Stated Preference Experiments" (bilingual), F. Angeli, Milan.

Danielis, R. et al.; (2005), "Logistics Managers" stated preferences concerning freight service attributes", Transport Research Part E 41, p. 201 – 215.

Dean, M.; Sebastia-Barriel, M.; (2004), "Why has world trade grown faster than output?", Bank of England Quarterly Bulletin, Vol. 44 Issue 3, p. 310 – 321.

Deardorff, A.V.; (1984), "Testing Trade Theories and Predicting Trade Flows", in R. W. Jones and P. B. Kenen, eds. "Handbook of International Economics", Vol. I, p. 467 – 517, Amsterdam, North Holland.

Deardorff, A.V.; (1995), "Determinants of bilateral Trade: Does Gravity work in a Neoclassical World?", Working Paper No. 5377, National Bureau of Economic Research, Cambridge, MA, USA.

Deno, K.T.; (1988), "The effect of public capital on U.S. manufacturing activity: 1970 to 1978", southern Economic Journal 55, 400 – 411.

Desmet, K.; Fafchamps, M.; (2003), "What are falling transport costs doing to spatial concentration across US counties?"; Working Paper No. 3853, Centre for Economic Policy Research, London, UK.

Dixit, A.; Stiglitz, J.E.; (1977), "Monopolistic Competition and Optimum Product Diversity", American Economic Review, No. 67 (3), p. 297 – 308.

Duffy-Deno, K.T.; Eberts, R.W.; (1989), "Public Infrastructure and Regional Economic Development: a simultaneous equation approach", Working Paper 8909, Federal Reserve Bank of Cleveland, USA.

Duggal, V.G. et al.; (1999), "Infrastructure and productivity: a nonlinear approach", Journal of Econometrics 92, 47 – 74.

Eberts, R.W.; (1986), "Estimating the Contribution of Urban Public Infrastructure to Regional Economic Growth", Working Paper no. 8610, Federal Reserve Bank of Cleveland.

Eberts, R.W.; Fogarty, M. S.; (1987), „Estimating the relationship between local public and private investment", Working Paper 8703, Federal Reserve Bank of Cleveland, USA.

Egger, P.; (2000), "A note on the proper econometric specification of the gravity equation", Economics Letters, Vol. 66, p. 25 – 31.
Egger, P.; (2001), "Transportation Costs in a Multilateral World – Assessing the Poolability of Gravity Data", Working Paper No. 183/2001, Österreichisches Institut für Wirtschaftsforschung, Vienna.
ECMT, (2003), "Time and Transport", Conclusions of Round Table 127, Paris, France.
European Commission; (2001); White Paper, "European transport policy for 2010: time to decide", Office for Official Publications of the European Communities, Luxembourg.
European Commission, Directorate General for Energy and Transport; (2003a), "European Engergy and Transport Trends to 2030".
European Commission, Directorate General, Joint Research Centre; (2003b); "Freight Transport Intensity of Production and Consumption", Report EUR 20864 EN.
Evenett, S.J.; Keller, W.; (1998), "On Theories explaining the Success of the Gravity Equation", National Bureau of Economic Research, Working Paper No. 6529, Cambridge, MA, USA.
Figuereido, O. et al.; (2002), "Home-field advantage: location decisions of Portuguese entrepreneurs", Journal of Urban Economics 52, 341 – 361.
Federal Highway Administration; (1998), "Contribution of Highway Capital to Output and Productivity Growth", FHWA, Washington, D. C.; http://www.fhwa.dot.gov/policy/gro 98ch2.htm
Feenstra, R.C.; (2002), "Border Effects and the Gravity Equation: Consistent Methods for Estimation", Scottish Journal of Political Economy, Vol. 49, No. 5, p. 491 – 505.
Field, A.; (2000), "Postgraduate Statistics: Cluster Analysis", Sussex University, http://www.sussex.ac.uk/Users/andyf/teaching/pg/cluster.pdf
Finger, J.M.; Yeats, A. J.; (1976), "Effective Protection by Transportation Costs and Tariffs: A Comparison of Magnitudes", Quarterly Journal of Economics, Vol. 90, No. 1, p. 169 – 176.
Forschungsgesellschaft für Strassen- und Verkehrswesen; (1997), "Empfehlungen für die Wirtschaftlichkeitsuntersuchungen auf Strassen, EWS", Band 132, FGSV Verlag, Köln
Frankel, J.A.; Romer, D.; (1999), „Does Trade cause Growth?", American Economic Review, Vol. 89 (3), p. 379 – 399
Gallup, J.L.; (1999), "Geography and Economic Development", International Regional Science Review, 22, p. 179 – 232.
Gao, T.; (2003), "Regional industrial growth: Evidence from Chinese industries", Regional Science and Urban Economics, Elsevier Science.
Garcia-Milà, T.; McGuire, T.J.; (1992), "The contribution of publicly provided inputs to states` economies", Regional Science and Urban Economics 22, 229 – 241.
Gaube, V. et al.; (2003), "Decoupling Economic Growth and Transport Demand", Intermediate Report, Joint Study carried out on behalf of Bundesministerium für Land- und Forstwirtschaft, Umwelt und Wasserwirtschaft, Vienna, Austria.

Gilbert, R.; Nadeau, K.; (2001); "Decoupling economic growth and transport demand: a requirement for sustainability"; Conference Paper for "Transportation and Economic Development 2002" of Transport Research Board.

Granger, C.W. J.; Newbold, P.; (1974), "Spurious Regressions in Econometrics", Journal of Econometrics, 2, p. 111 - 120.

Green, P.E.; (1984), "Hybrid Models for Conjoint Analysis: An Expository Review", Journal of Marketing Research, 21, p. 155 - 169.

Green, P.E. et al.; (2001), "Thirty Years of Conjoint Analysis: Reflections and Prospects", Interfaces, 31:3, Part 2 of 2, May/June 2001, p. 56 - 73

Griffiths, W.E. et al.; (1992), "Learning and Practicing Econometrics", John Wiley & Sons Inc.

Guild, R.L., (1998), "Infrastructure investment and regional development: theory and evidence", Department of Planning Working Paper Series, University of Auckland.

Halver, W.; (2003), "Methoden der Standortanalyse", Halver Research Standortforschung Wirtschaftsgeographie, Nordrhein-Westfalen.

Harrigan, J.; (1993), "OECD Imports and Trade Barriers in 1983", Journal of International Economics, vol. 35 n. 1 - 2, p. 91 - 111.

Haughwout, A.F.; (1998), "Aggregate Production Functions, Interregional Equilibrium, and the Measurement of Infrastructure Productivity", Journal of Urban Economics 44, 216 - 227.

Haughwout, A.F.; (2002), "Public infrastructure investments, productivity and welfare in fixed geographic areas", Journal of Public Economics 83, 405 - 428.

Haveman, J.; Hummels, D.; (2004), "Alternative hypotheses and the volume of trade: the gravity equation and the extent of specialization", Canadian Economics Association.

Head, K. et al.; (1995), "Agglomeration Benefits and Location Choice: Evidence form Japanese Manufacturing Investments in the United States", Journal of International Economics 38, p. 223 - 247.

Head, K.; (2000), "Gravity for Beginners", Faculty of Commerce, University of British Columbia, Vancouver, Canada.

Helbing, D.; (1997), "Verkehrsdynamik", Springer, Berlin, in Gössel, F.; (2005), "Informationsentropische, spektrale und statistische Untersuchungen fahrzeuggenerierter Verkehrsdaten unter besonderer Berücksichtigung der Auswertung und Dimensionierung von FCD-Systemen", Dissertation, Fakultät für Verkehrswissenschaften, TU Dresden.

Holl, A.; (2003), "Manufacturing location and impacts of road transport infrastructure: Empirical evidence from Spain", Department of Town and Regional Planning, University of Sheffield.

Holtz-Eakin, D.; (1994), "Public Sector Capital and the Productivity Puzzle", Review of Economics and Statistics 76, p. 12 - 21.

Holtz-Eakin, D.; Schwartz, A. E.; (1995), „Infrastructure in a structural model of economic growth", Regional Science and Urban Economics 25, 131 - 151.

Holtz-Eakin, D.; Lovely, M. E.; (1996), "Scale economies, returns to variety, and the productivity of public infrastructure", Regional Science and Urban Economics 26, 105 - 123.

Hulten, C.R.; Schwab, R. M.; (1984), „Regional productivity growth in U.S. manufacturing: 1951 - 1978", American Economic Review 74, 152 - 162.

Hulten, C.R.; Schwab, R. M.; (1991), „Pubic capital formation and the growth of regional manufacturing industries", National Tax Journal 43, 121 - 143.

Hummels, D.; (1998a), "Vertical specialisation and the changing nature of world trade", Economic Policy Review, Vol. 4 (2), Federal Reserve Bank of New York, USA.

Hummels, D.; (1998b), "Towards A Geography of Transport Costs", mimeo University of Chicago.

Hummels, D. et al.; (1999a), "The Nature and Growth of Vertical Specialisation in World Trade", University of Chicago.

Hummels, D.; (1999b), "Have International Transport Costs Declined", mimeo University of Chicago.

Hummels, D.; (2001), "Time as a trade barrier", Working Paper, Purdue University, West Lafayette, Indiana.

Ibrahim, I.; (2002), "On Exports and Economic Growth", Jurnal Pengurusan 21 (2002), p. 3 - 18.

Irwin, D.A.; Trevio, M.; (2000), "Does Trade raise Income: Evidence From The Twentieth Century", Working Paper No. 7745, National Bureau of Economic Research, Cambrigde, MA.

Ishii, J.; Yi, K.-M.; (1997), "The growth of world trade", Federal Reserve Bank of New York, USA.

Islam, M.N.; (1998), "Export expansion and economic growth: testing for cointegration and causality", Applied Economics, 1998, 30, p. 415 - 425.

Kerner, B.; (2000), "Theory of Breakdown Phenomenon at Highway Bottlenecks", Paper No. 00-1573, Transportation Research Board, 79th Annual Meeting, Washington D.C. 2000 in "Einfluss von Streckenbeeinflussungsanlagen auf die Kapazität von Autobahnabschnitten sowie die Stabilität des Verkehrsflusses", Veröffentlichungen aus dem Institut für Strassen- und Verkehrswesen, Heft 35, Universität Stuttgart.

Kim, J.H.; (2003), "An empirical analysis on public capital and aggregate output", Research Proposal, University of Missouri-Columbia.

Kim, M.K. et al.; (2003), "Determining Bilateral Trade Patterns Using a Dynamic Gravity Equation", Agribusiness & Applied Economics Report No. 525, North Dakota State University, Fargo, USA.

Kim, Y.; Keller, H.; (2001), "Zur Dynamik von Verkehrszuständen im Fundamentaldiagramm", Straßenverkehrstechnik 9/2001, p. 433 - 442 in Schick, P.; (2003), "Einfluss von Streckenbeeinflussungsanlagen auf die Kapazität von Autobahnabschnitten sowie die Stabilität des Verkehrsflusses", Veröffentlichungen aus dem Institut für Strassen- und Verkehrswesen, Heft 35, Universität Stuttgart.

Krugman, P.; (1980), "Scale Economies, Product Differentiation, and the Pattern of Trade", American Economic Review, Vol. 70, p. 950 - 959.

Krugman, P.; (1991), "Increasing Returns and Economic Geography", Journal of Political Economy 99, p. 483 – 499.
Krugman, P.; Obstfeld, M.; (2001), "International Economics", 5th edition, Addison-Wesley.
Krugman, P.; Venables, A. J.; (1995), "The Seamless World: A Spatial Model of International Specialization", Discussion Paper No. 1230, Centre for Economic Policy Research, London, UK.
Kruskal, J.B.; (1965), "Analysis of Factorial Experiments by Estimating Monotone Transformations of the Data", Journal of the Royal Statistical Society, Series B, Vol. 27, p. 251 – 263.
Kumar, N.; (2001), "Infrastructure availability, foreign direct investment flows and their export orientation: a cross-country exploration", Research and Information System for Developing Countries, New Delhi.
Lai, H.; Trefler, D.; (2002), "The Gains from Trade with Monopolistic Competition: Specification, Estimation, and Mis-Specification", Working Paper 9169, National Bureau of Economic Research, Cambridge, MA, USA.
Lakshmanan, T.R.; Anderson, W. P.; (2002), "Transportation Infrastructure, Freight Services Sector and Economic Growth", White Paper, The U. S. Department of Transportation, Federal Highway Administration, USA.
Laursen, K.; (2000), "Trade, Specialisation, Technology and Economic Growth", Edward Elgar Publishing, Cheltenham, UK.
Leamer, E.; Levinsohn, J.; (1997), "International Trade Theory, the Evidence", in Handbook of International Economics, ed. Grossman and Rogoff Vol. 3, Amsterdam, North-Holland.
Louviere, J. J. et al.; (2000), "Stated Choice Methods", Cambridge University Press, UK.
Luce, R.D.; Tukey, J. W.; (1964), "Simultaneous Conjoint Measurement: A New Type of Fundamental Measurement", Journal of Mathematical Psychology, Vol. 1, p. 1 – 27.
Lynde, C.; Richmond, J.; (1992), "The role of public capital in production", Review of Economics and Statistics 74, 37 – 44.
Lynde, C.; Richmond, J.; (1993), "Public capital and long-run costs in U.K. manufacturing", The Economic Journal 103, 880 – 893.
Maffei, M.J.; Meredith, J.; (1995), "Infrastructure and flexible manufacturing technology: Theory development", Journal of Operations Management 13, 273 – 298.
Maier et al.; (2002), "Modelling Preferences and Stability among Transport Alternatives", Transportation Research Part E, vol. 38.
Markusen, J. R.; (1995), "International Trade", McGraw Hill, New York.
Marshall, A.; (1920), "Principles of Economics", 8th edition, Mcmillan, London, UK.
Martin, P.; Rogers, C.A.; (1995), "Industrial location and public infrastructure", Journal of International Economics 39, 335 – 351.
Mazzenga, E.; Ravn, M.; (2002), "International Business Cycles: The Quantitative Role of Transportation Costs", Working Paper No. 3530, Centre for Economic Policy Research, London, UK.

Melitz, M.J.; (2002), "The Impact of Trade on Intra-Industry Reallocations and Aggregate Industry Productivity", NBER Working Papers No. 8881.

Micco, A.; Pérez, N.; (2001), "Maritime Transport Costs and Port Efficiency", Seminar Paper for Seminar "Towards Competitiveness: An Institutional Path", Inter-American Development Bank, Santiago, Chile.

Ministry of Transport and Communications, Finland; (2002); "Transport Infrastructure 2030 – Meeting the Challenges of Concentrating Population and Industrial Changes", Programmes and Strategies 3/2002, MTC, Finland.

Morrison, C.J.; Schwartz, A. E.; (1992), „State infrastructure and productive performance", Working Paper No. 3981, National Bureau of Economic Research, Cambridge, MA.

Munnell, A.H.; (1990a), "Why has productivity growth declined? Productivity and Public Investment", New England Economic Review, Sept./Oct., 11 – 32.

Munnell, A.H.; (1990b), "How does public infrastructure affect regional economic performance?", New England Economic Review, Sept./Oct., 11 – 32.

Munnell, A.H.; (1992), "Infrastructure Investment and Economic Growth", Journal of Economic Perspectives, American Economic Association, vol. 6 (4), p. 189 – 198.

Nadiri, M.I.; Mamuneas, T. P.; (1994), "The effects of public infrastructure and R & D capital on the cost structure and performance of U.S. manufacturing industries", The Review of Economics and Statistics 76 (1), 22 – 37.

Nicolini, R.; (2000), "Local agglomerations and trade: an empirical investigation", Département des Sciences Ecoomiques, Université de Catholique de Louvain, France.

Nijkamp, P.; Reggiani, A.; Bolis, S.; (1997), "European freight transport and the environment: empirical applications and scenarios".

Nijkamp, P.; van Delft, H.; van Veen-Groot, D.; (1998), "Sustainable mobility and globalisation: new research and policy challenges", Research Memoranda, Faculty of Economics, University of Amsterdam.

Noguer, M.; Siscart, M.; (2005), "Trade raises Income: A Precise and Robust Result", Journal of International Economics, Vol. 65 (2), p. 447 – 460.

OECD; (2003a), "Qualitative assessment of the benefits of trade facilitation", TD/TC/WP 31.

OECD; (2003b), "Project on Decoupling Transport Impacts and Economic Growth", Working Paper on National Environmental Policy, Working Group on Transport.

Ottaviano, G.I.P.; Thisse, J.-F.; (2003), "Agglomeration and economic theory", Discussion Paper No. 3838, Centre for Economic Policy Research, London.

Overman, H.G. et al.; (2001), "The Economic Geography of Trade, Production and Income: A Survey of Empirics", Centre for Economic Performance, London School of Economics, UK.

Pinna, A.; (1993), "Sectoral Composition of Trade and Economic Growth: Some New Robust Evidence", Department of Economics, University of Warwick.

Pöyhönen, P.; (1963), "A Tentative Model for the Volume of Trade between Countries", Hamburger Weltwirtschaftliches Archiv 90, p. 93 – 100.

Rodriguez, F.; Rodrik, D.; (1999), "Trade Policy and Economic Growth: A Sceptic's Guide to Cross-National Evidence", NBER Working Paper No. 7081.

Rossi-Hansberg, E.; (2003), "A Spatial Theory of Trade", Stanford University.

Rovolis, A.; Spence, N.; (2002), "Duality theory and cost function analysis in a regional context: the impact of public infrastructure capital in the Greek regions", The Annals of Regional Science 36, 55 – 78.

Sampson, G.P.; Yeats, A. J.; (1977), "Tariff and Transport Barriers Facing Australian Exports," Journal of Transport Economics and Policy, Vol. 11, No. 2, p. 141 – 54.

Sawtooth Software Inc.; (2002), "ACA User Manual vs. 5"

Schick, P.; (2003), "Einfluss von Streckenbeeinflussungsanlagen auf die Kapazität von Autobahnabschnitten sowie die Stabilität des Verkehrsflusses", Veröffentlichungen aus dem Institut für Strassen- und Verkehrswesen, Heft 35, Universität Stuttgart.

Skiera, B.; Gensler S.; (2002); „Berechnung von Nutzenfunktionen und Marktsimulationen mit Hilfe der Conjoint-Analyse Teil 1", Wirtschaftswissenschaftliches Studium, Heft 4.

Solow, R.M.; (1956), "A contribution to the theory of economic growth", Quarterly Journal of Economics 70, p. 65 – 94.

Song, L.L.; (2002), "Public capital, congestion and private production in Australia", Melbourne Institute Working Paper No. 23/02.

Stephan, A.; (1997), "The Impact of Road Infrastructure on Productivity and Growth: Some Preliminary Results for the German Manufacturing Sector", Discussion Paper FS IV 97 – 47, Wissenschaftszentrum Berlin.

Steininger, K.W.; (2001), "International Trade and Transport", Edward Elgar Publishing, Cheltenham, UK.

Stiglitz, J.; (1988), "Economics of the public sector", Norton, New York.

Stijns, J.P.; (2003), "An Empirical Test of the Dutch Disease Hypothesis using a Gravity Model of Trade", Draft for the Congress of EEA, Stockholm, Sweden.

Sturm, J.E.; de Haan, J.; (1995), „Is public expenditure really productive?", Economic Modelling 12 (1), 60 – 72.

Tinbergen, J.; (1962), "Shaping the World Economy. Suggestions for an International Economic Policy", The Twentieth Century Fund, New York, USA.

Venables, A.J.; Limao, N.; (2001), "Infrastructure, Geographical Disadvantage, Transport Costs, and Trade", The World Bank Economic Review, Vol. 15, No. 3, 451 – 479.

Venables, A.J.; Limao, N.; (2002), "Geographical disadvantage: a Heckscher-Ohlin-von-Thünen model of international specialisation", Journal of International Economics 58, p. 239 - 263.

Vickerman, R.; (1995), "Location, accessibility and regional development: the appraisal of trans-European networks", Transport Policy, Vol. 2, No. 4, p. 225 - 234, Elsevier Science, UK.

Wälde, K.; Wood, C; (2004), "The Empirics of Trade and Growth: Where are the Policy Recommendations?", Economic Papers, No. 204, European Commission, Directorate-General for Economic and Financial Affairs, Brussels.

Wall, H.J.; (1999), "Using the Gravity Model to Estimate the Costs of Protection", Federal Reserve Bank of St. Louis, USA.

Waters, W.; (1970), "Transport Costs, Tariffs, and the Pattern of Industrial Protection", American Economic Review, 60, p. 1013 - 1020.

WTO; (2007);http://www.wto.org/English/thewto_e/minist_e/min99_e/english/about_e/22fact_e.htm

Yanikkaya, H.; (2002), "Trade Openness and Economic Growth: a Cross-Country Empirical Investigation", Journal of Development Economics 72, p. 57 - 89.

Yeaple, S.R.; (2002), "A Simple Model of Firm Heterogeneity, International Trade and Wages", University of Pennsylvania mimeo.

Yi, K.-M.; (2003), "Can Vertical Specialization Explain the Growth of World Trade?", Journal of Political Economy, Vol. 111, No. 1, p. 52 - 102.

Young, F.W.; (1969), "Polynomial Conjoint Analysis of Similarities: Definitions for a Special Algorithm", Research Paper No. 76, Chapel Hill: University of North Carolina, Psychometric Laboratory.

Zegeye, A.A.; (2000), "U. S. Public infrastructure and its contribution to private sector productivity", Working Paper 329, U. S. Bureau of Labor Statistics.

MIX
Papier aus verantwortungsvollen Quellen
Paper from responsible sources
FSC® C105338

If you have any concerns about our products,
you can contact us on
ProductSafety@springernature.com

In case Publisher is established outside the EU,
the EU authorized representative is:
Springer Nature Customer Service Center GmbH
Europaplatz 3, 69115 Heidelberg, Germany

Printed by Libri Plureos GmbH
in Hamburg, Germany